T0221525

ROCK SCOUR DUE TO FALLING HIGH-VELOCITY JETS

Co-hosters

ÉCOLE POLYTECHNIQUE
FÉDÉRALE DE LAUSANNE

SWISS COMMITTEE ON DAMS

GEO
INSTITUTE

of the American Society of Civil Engineers

ASCE

American Society of Civil Engineers

SCHWEIZERISCHER WASSERWIRTSCHAFTSVERBAND

Sponsors

ELECTROWATT-EKONO
Jaakko Pöyry Group

 LOMBARDI
ENGINEERING LIMITED

Engineering
&
Hydrosystems Inc

INGEGNERIA MAGGIA SA
INGEGNERI CONSULENTI
6601 LOCARNO 1

STUCKY   Ingénieurs-conseils SA

PROCEEDINGS OF THE INTERNATIONAL WORKSHOP ON ROCK SCOUR DUE TO HIGH-VELOCITY JETS, LAUSANNE, SWITZERLAND, 25-28 SEPTEMBER 2002

Rock Scour
due to falling high-velocity jets

Edited by
Anton J. Schleiss & Erik Bollaert
Laboratory of Hydraulic Constructions, EPFL, Lausanne, Switzerland

A.A. BALKEMA PUBLISHERS LISSE / ABINGDON / EXTON (PA) / TOKYO

Published by: A.A. Balkema, a member of Swets & Zeitlinger Publishers
www.balkema.nl and www.szp.swets.nl

ISBN 90 5809 518 5

Printed in the Netherlands

Rock Scour due to falling High-velocity Jets - Schleiss & Bollaert (eds)
© 2002 Swets & Zeitlinger, Lisse, ISBN 90 5809 518 5

Table of contents

Preface   IX

Organization   XI

Introductory lecture

Scour evaluation in space and time – the challenge of dam designers   3
A.J. Schleiss

Case studies and prototype observations of scour

Review of plunge pool rock scour downstream of Srisailam Dam   25
P.J. Mason

Scour hole geometry for Fort Peck Spillway   33
A.F. Babb, D. Burkholder & R.A. Hokenson

Slab stability in hydraulic jump stilling basins as derived from physical   43
modeling: the Mignano Dam case
E. Caroni, V. Fiorotto & M. Belicchi

Local rock scour downstream large dams   55
R.A. Lopardo, M.C. Lopardo & J.M. Casado

Quantification of extent of scour using the Erodibility Index Method   63
G.W. Annandale

A review on physical models of scour holes below large dams in Iran   73
J. Attari, F. Arefi & F. Golzari

Physical modeling and scale effects

Drag of emergent and submerged rectangular obstacles in turbulent   83
flow above bedrock surface
P.A. Carling, M. Hoffmann, A.S. Blatter & A. Dittrich

Parametric analysis of the ultimate scour and mean dynamic pressures   95
at plunge pools
L.G.E. Castillo

Pressure fields due to the impingement of free falling jets on a riverbed   105
J. Puertas & J. Dolz

Jet aeration and jet air entrainment in plunge pools

Does an aerated water jet reduce plunge pool scour?   117
H.-E. Minor, W.H. Hager & S. Canepa

Reduction of plunge pool floor dynamic pressure due to jet air entrainment   125
J.F. Melo

The influence of plunge pool air entrainment on the presence of free   137
air in rock joints
E. Bollaert

*Time-scale effects, break-up resistance and hydraulic-mechanical
interaction*

The Erodibility Index Method: an overview   153
G.W. Annandale

A physically-based engineering model for the evaluation of the   161
ultimate scour depth due to high-velocity jet impact
E. Bollaert & A.J. Schleiss

Geotechnical aspects of rock erosion   175
J.H. May, J.L. Wibowo & C.C. Mathewson

Scour of rock due to high-velocity jet impact: a physically-based scour   187
model compared to Annandale's Erodibility Method
E. Bollaert, G.W. Annandale & A.J. Schleiss

Quantification of the relative ability of rock to resist scour   201
G.W. Annandale

Wave impact induced erosion of rock cliffs   213
G. Wolters & G. Müller

Numerical modeling of rock scour

Dynamic response of the drainage system of a cracked plunge pool liner   227
due to free falling jet impact
M. Mahzari, F. Arefi & A.J. Schleiss

Genetic algorithm optimization of transient two-phase water pressures   239
inside closed-end rock joints
E. Bollaert, S. Erpicum, M. Pirotton & A.J. Schleiss

History of energy dissipators

Short history of energy dissipators in hydraulic engineering   253
W.H. Hager

Author Index   263

Preface

Scour of rock foundations due to plunging high-velocity jets is a well-known phenomenon occurring at hydraulic outlet structures in general and at spillways of large dams in particular. Since scour formation downstream of hydraulic structures and dams is a safety issue, the designers have to estimate at least the ultimate scour depth. Since long time this analysis is performed by means of empirical expressions, mostly developed from physical model tests and prototype observations. However, these approaches cannot consider the full complexity of the hydrodynamic-geotechnical scour problem. Therefore, many researchers and engineers tried in the last decades to develop physically better-based evaluation methods. With the aim to focus on these recently performed and ongoing research as well as novel practical design experience, a workshop was organized bringing together the leading scientists and the specialized engineers in the field of scour assessment. The major aspects of rock scour were discussed, such as:
-   Physical modeling and scale effects,
-   Prototype observations,
-   Time-scale effects and break-up resistance of rock,
-   Hydraulic – mechanical interactions,
-   Jet aeration and air entrainment in plunge pools,
-   Numerical modeling,
-   Protection measures.

Furthermore, the presentation of interesting case studies allowed the exchange of ideas between researchers and practicing engineers.

This book contains the papers presented at the "International Workshop on Rock Scour" held in September 2002 at the Swiss Federal Institute of Technology in Lausanne (EPFL), Switzerland. It gives an overview of the state-of-the-art and the trends of the ongoing research, and it highlights the need for future investigations.

The editors gratefully acknowledge the co-hosting of the International Association of Hydraulic Engineering and Research (IAHR), the American Society of Civil Engineers (ASCE), the US Army Corps of Engineers (USACE), the Geo-Institute of the ASCE, the Swiss Committee on Dams (SWISSCOD) and the Swiss Association of Water Resources (SWV), as well as the support of members of the scientific committee. We appreciate the financial support of Electrowatt-Ekono Ltd., Switzerland; Engineering & Hydrosystems Inc., United States; Lombardi Engineering Ltd., Switzerland; IM Maggia Engineering Ltd., Switzerland and Stucky Consulting Engineers Ltd., Switzerland.

The organization of the symposium and the preparation of the book at hand were carried out by the Laboratory of Hydraulic Constructions (LCH) of the EPFL.

Anton J. Schleiss & Erik Bollaert
Lausanne, September 2002

Organization

Introductory lecture

Scour evaluation in space and time – the challenge of dam designers

A.J. Schleiss
Laboratory of Hydraulic Constructions (LCH), Swiss Federal Institute of Technology Lausanne (EPFL), Lausanne, Switzerland

ABSTRACT: The main questions arising during a dam project in view of safety against scour formation downstream of spillways and bottom outlets are discussed. First the physical processes involved in scour formation are analyzed based on the state-of-the-art of the knowledge and understanding. Then the different scour evaluation methods from simple empirical formula to complex scour models are briefly presented. The main difficulties encountered when estimating scour depth are discussed, such as the choice of the appropriate theory, interpretation of hydraulic model tests and prototype observations, scour rate and prevailing discharge. Furthermore the selection of the return period of the flood event is discussed, for which the scour formation and the control measures have to be evaluated. Finally the options of measures for scour control are presented and discussed.

1 INTRODUCTION

The safety of dams during flood events has to be ensured by an appropriate capacity of the releasing structures. These spillways are designed for the so-called design or project flood, which is defined according to the legislation of the different countries, typically a 1000-years flood. Furthermore, normally it is requested, that the dam should not be endangered for even higher floods, the so-called safety check flood, which corresponds typically to a flood between a 10'000-years flood and a PMF.

Depending on the type of spillway, high velocity jets can occur which are guided by the releasing structures into the tailwater at a certain distance from the dam. At the zone of impact of these high-energy jets, the riverbed will be scoured. Since scour due to plunging jets can reach considerable depth even in rocky river beds, instability of the valley slopes has to be feared, which can endanger in some cases the foundation and the abutment of the dam itself. Such scour problems occur especially at dams were the spillways are combined with the dam structure itself and, consequently, the impact zone of the high-velocity falling jets is relatively close to the dam. This is typically the case with concrete dams, where high velocity falling jets can be created by gated or ungated crest spillways (arch dams only), chute spillways followed by a ski jump and orifice spillways as well as high capacity bottom outlets. Severe scour conditions occur especially in the case of high concrete arch dams in narrow valleys with high flood discharges.

Such a typical spillway arrangement is shown in Figure 1 and Figure 2 at the example of Khersan III Dam Project in Iran. Flood handling at the 175 m high double arch dam will be provided by three separate spillway facilities:

- A two-bay chute flip bucket spillway on the right abutment with an ogee crest at El. 1404.5 m controlled by 11.5 m wide by 13.5 m high radial gates (capacity of 4240 m^3/s at PMF flood level El. 1426.3 m)

- An uncontrolled crest spillway with ogee crest divided in 6 bays; two 13.5 m and two 19.5 m wide bays at El. 1418 m and two 12.5 m wide bays at El. 1421 m (total capacity of 3360 m³/s at PMF flood level El. 1426.3 m).
- Two bottom outlets with centerline at El. 1330 and 1345 m with service gate openings 3 m wide by 4 m high (total capacity of 395 m³/s at PMF flood level El. 1426.3 m)

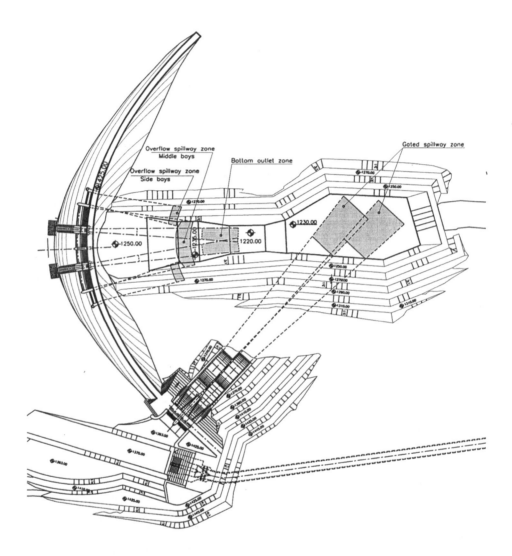

Figure 1. Khersan III Dam Project in Iran. Layout of arch dam with spillway structures and its jet impact zones.

In today's spillway design of dams there is a tendency of increasing the unit discharge of the high velocity jet leaving the appurtenant structures. In gated chute flip bucket spillway, unit discharges in the range of 200 to 300 m³/sm are not rare anymore, since cavitation risk in chutes can be mitigated by the help of bottom aerators. Uncontrolled crest spillways for arch dams are designed nowadays for specific discharges up to 70 m³/sm and up to 120 m³/sm by installing gates on the crest. With the latest high pressure gate technology, low level orifice spillways can evacuate unit discharges in the range of 300 to 400 m³/sm.

4

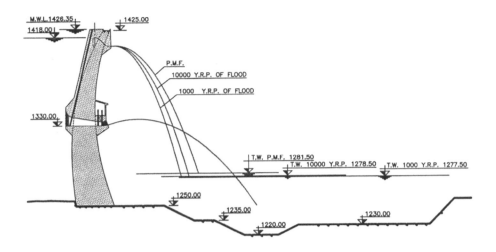

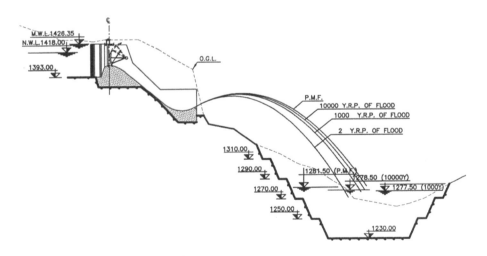

Figure 2. Khersan III Dam Project in Iran. Longitudinal profiles through crest spillway and gated side spillway and plunge pools.

Therefore in many dam projects it is a challenge for dam designers to answer the following questions:
- What will be the evolution and extent of scour downstream of the dam at the jet impact zone?
- Are the stability of the valley slopes and the foundation of the dam itself endangered?
- Is a tailpond dam required to create a water cushion and how does it affect the scour depth?
- Is a pre-excavation of the rocky river bed required and/or has the plunge pool to be lined?
- Is the powerhouse operation influenced by scour formation?

2 THE SCOUR PROCESS

2.1 *The physical processes*

Scouring is a complex problem and has been studied since longtime. As illustrated in Figure 3, scour can be described by a series of physical processes as (Bollaert 2002)

a) free falling jet behavior in the air and aerated jet impingement
b) plunging jet behavior and turbulent flow in the plunge pool
c) pressure fluctuation at the water-rock interface
d) propagation of dynamic water pressures into rock joints
e) hydrodynamic fracturing of closed end rock joints and splitting of rock in rock blocks
f) ejection of the so formed rock blocks by dynamic uplift into the plunge pool
g) break-up of the rock blocks by the ball milling effect of the turbulent flow in the plunge pool
h) formation of a downstream mound and displacement of the scoured materials by sediment transport

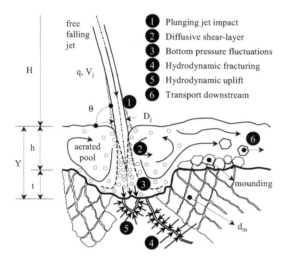

Figure 3. Main parameters and physical-mechanical processes involved in the scouring phenomenon (according Bollaert & Schleiss 2001c)

2.2 *Jet behavior in the air*

In evaluating the scour caused by free jets, it is at first necessary to predict the jet trajectory so that the location of the jet impingement in the plunge pool and the zone of the scour hole are known (Whittaker & Schleiss 1984). The behavior of an ideal jet can be easily assessed by using ballistic equations. Nevertheless, at prototype jets effects as air drag, disintegration of the jet in the air and initial flow aeration in long chutes have to be considered. A number of researchers have developed equations to predict the jet trajectory, i.e. the jet travel length accounting for these effects (Gun'ko et al. 1965, Kamenev 1966, Kawakami 1973, Zvorykin et al. 1975, Taraimovich 1980, Martins 1977).

The spread of the jet during fall is also an issue and has been addressed on an empirical basis by Taraimovich 1980 and U.S.B.R. 1978. More recently, the lateral spread has been related to its initial turbulence intensity (Ervine & Falvey, 1987; Ervine et al. 1997). Typical angles of jet spread are 3 to 4 % for roughly turbulent jets. Turbulence intensity for free overfall jets is less than 3 %, for flip bucket jets between 3 and 5 % and for orifice jets between 3 and 8 % (Bollaert 2002).

6

2.3 Jet behavior in the plunge pool and pressure fluctuations

As the jet plunges into the pool, a considerable amount of air is entrained, corresponding to an air concentration of 40 to 60% at typical jet velocities of 30 m/s at impingement (Bollaert 2002). Several expressions defining the air content at the point of impact of the jet in the plunge pool have been developed by scale model tests. Some of them can be reasonably extended towards prototype velocities (Van de Sande 1973, Bin 1984, Ervine et al. 1980, Ervine 1998).

The flow conditions in the plunge pool can be characterized by a high-velocity, two-phase turbulent shear layer flow and a macroturbulent flow outside of this zone (Bollaert 2002). The shear layer produces severe pressure fluctuations at the water-rock interface and is highly influenced by aeration. Existing theories on two-dimensional vertical jet diffusion in a semi-infinite or bounded medium define the outer limits of the water-rock interface that is directly subjected to these pressures. The diffusion of 2D-jet has been investigated by many researchers (Bollaert & Schleiss 2002a). The concept of a jet of uniform velocity field penetrating into a stagnant fluid is based on the progressive growing of the thickness of the related boundary layer by exchange of momentum. In this shear layer the total cross section of the jet increases whereas the non-viscous core of the jet decreases. The inner angle of diffusion is about 8° for highly turbulent impinging jets. When the plunge pool is deep enough, the core jet disappears and a fully developed jet occurs. The angle of diffusion of the jet through the pool depends on the degree of turbulence and aeration of the jet at impact and can be estimated at around 15° (Ervine & Falvey 1987). The most severe hydrodynamic action on the plunge pool bottom occurs in the impingement region, where the hydrostatic pressure of the free jet region is progressively transformed into fluctuating high stagnation pressures and into an important bottom shear stress.

Knowledge on the statistical characteristics and of the spatial distribution of the pressure fluctuations has been acquired by physical modeling of hydraulic jumps and plunging jets (Tos & Bowers 1988, Ballio et al. 1992, Bollaert & Schleiss 2002b). The pressure patterns generated by core jet impact are generally quite constant with high values in the core and much lower values directly outside the core. The pressures for developed jet impact are completely different from pressures generated by core jet impact. Due to the high turbulence level and the two phase character of the shear layer, developed jet impact conditions can generate much more severe dynamic pressures at the pool bottom. Therefore, not every water cushion has a retarding effect on the scour formation. Power spectral content curves indicate that core jet impact creates quasi-linear decaying spectra, even at very high frequencies. The rate of energy decrease follows f^{-1} (Bollaert & Schleiss 2001a). On the other side developed jet impact shows two spectral ranges: a low frequency part (up to 100 Hz) where a considerable amount of energy exists and a high frequency part (> 100 Hz) with a rate of energy decay according $f^{-7/3}$.

It may be concluded that the spectral energy content of a high velocity plunging jet extends over a much wider frequency range than what is generally assumed for macroturbulent flows in plunge pools (up to 25 Hz).

2.4 Propagation of dynamic water pressures into rock joints, hydrodynamic fracturing and dynamic uplift

The transfer of pool bottom pressures into rock joints results in transient flow that is governed by the propagation of pressure waves. For closed-end rock joints, as encountered in a partially jointed rock mass, the reflection and superposition of pressure waves generate a hydrodynamic loading at the tip of the joint. If the corresponding stresses at the tip of the joint exceed the fracture toughness and the initial compressive stresses in the rock, the rock will crack and the joint can further grow. In the case of open-end rock joints in a fully jointed rock mass, the pressure waves inside the joints create a significant dynamic uplift force on the rock blocks. This dynamic uplift force will break up the remaining rock bridges in the joints by fatigue and, if high enough, eject the so formed rock blocks from the rock mass into the macroturbulent plunge pool flow.

The presence of such pressure waves in rock joints and significant amplifications due to resonance phenomena could be observed and measured with an experimental study at near prototype scale (Bollaert 2002, Bollaert & Schleiss 2002b).

2.5 Ball milling effect of the turbulent flow in the plunge pool and formation of a downstream mound

Once the rock blocks are formed and ejected by the dynamic uplift from the surrounding rock mass into the pool, they can be taken up by the macroturbulent eddies. If the block is too big to be transported by the flow, it will be broken-up after some time by the ball milling effect of the eddies in the plunge pool. Having attained the required minimum size, the rock blocks will be displaced downstream by the flow and deposited on the mound or carried away by sediment transport in the river. The mound may limit the depth of scour but also raise the tailwater level. Exceeding a critical level, the increased tailwater may interfere with the operation of bottom outlets or reduce the net head of the powerhouse, if it is located upstream of the mound. If the mound does limit the depth of the scour, the scour is considered to have attained a so called dynamic limit. However if the mound is removed and the scour proceeds to a maximum possible extent, it is considered to have attained the ultimate static limit (Eggenberger & Müller 1944).

3 SCOUR EVALUATION METHODS

3.1 General overview

The existing scour evaluation methods can be grouped as follows (Bollaert & Schleiss 2002a):

- Empirical approaches based on laboratory and field observations
- Analytical-empirical methods combining laboratory and field observations with some physics
- Approaches based on extreme values of fluctuating pressures at the plunge pool bottom
- Techniques based on time-mean and instantaneous pressure differences and accounting for rock characteristics
- Scour model based on fully transient water pressures in rock joints

The most common methods used for scour evaluation due to falling, high velocity jets are listed in Table 1 together with the considered physical parameters. The parameters can be related to the three phases water, rock and air, which are involved in the scour process. Time evolution is a further parameter.

3.2 Empirical formulae

A large number of empirical equations have been developed for predicting the scour from plunging jets. These empirical formula were mostly derived from hydraulic model tests in the laboratory but also from prototype observation and are widely used in practice for design purposes. Some expressions are of general applicability, others are specific to free overfall jets, ski-jumps or orifice spillways. A comprehensive overview and comparison of most of the known formula have been given by Whittaker & Schleiss (1984), Mason & Arumugam (1985) and Bollaert (2002).

Type	Year	Author(s)	Applicability	Time	Hydraulic characteristics									Geomechanical characteristics									Aeration characteristics			
					hydrostatic						hydrodynamic			gran. soil			jointed rock mass									
				T	q	h	H	g	V_j	θ	RMS	Sxx(f)	trans	d_m	w_a	$ρ_s$	$σ_c,σ_t$	RQD	N_j	z	α	$φ_i$	C	β	Tu	L_b
				[-]	[m²/s]	[m]	[m]	[m/s²]	[m/s]	[°]	[-]	[m²]	[-]	[mm]	[m/s]	[kg/m³]	[N/m²]	[%]	[-]	[-]	[°]	[°]	[-]	[-]	[%]	[m]
empirical	1932	Schoklitsch	plunging jet	-	■	-	■	-	-	-	-	-	-	■				-	-	-	-	-	-	-	-	-
	1937	Veronese A	horiz. & plunging jet		■	-	■	-	-	-	-	-	-	■				-	-	-	-	-	-	-	-	-
	1937	Veronese B	as A, but d_m < 0.005m		■	-	■	-	-	-	-	-	-	-				-	-	-	-	-	-	-	-	-
	1939	Jaeger	plunging jet	-	■	■	■	-	-	-	-	-	-	■				-	-	-	-	-	-	-	-	-
	1953	Doddiah et al.	plunging jet	■	■	-	■	-	-	-	-	-	-	■	■			-	-	-	-	-	-	-	-	-
	1957	Hartung	plunging jet	-	■	-	■	■	-	-	-	-	-	■				-	-	-	-	-	-	-	-	-
	1963	Rubinstein	ski-jump, rock cubes	-	-	■	■	■	■	■	-	-	-	-				-	-	-	■	-	■	-	-	-
	1966	Damle et al.*	ski-jump	-	■	-	■	-	-	-	-	-	-	-				-	-	-	-	-	-	-	-	-
	1967	Kotoulas	plunging jet	-	■	-	■	-	-	-	-	-	-	■				-	-	-	-	-	-	-	-	-
	1969	Chee & Padiyar	flip bucket	-	■	-	■	-	-	-	-	-	-	■				-	-	-	-	-	-	-	-	-
	1974	Chee & Kung	plunging jet	-	■	-	■	■	-	■	-	-	-	■				-	-	-	-	-	-	-	-	-
	1973	Martins A	plunging jet, rock cubes	-	■	-	■	-	-	-	-	-	-	-				-	-	-	■	-	-	-	-	-
	1975	Martins B	ski-jump	-	■	-	■	-	-	-	-	-	-	-				-	-	-	-	-	-	-	-	-
	1978	Taraimovich	ski-jump	-	-	-	■	-	■	■	-	-	-	-	■			-	-	-	-	-	-	-	-	-
	1981	INCYTH	plunging jet	-	■	-	■	-	-	-	-	-	-	-				-	-	-	-	-	-	-	-	-
	1982	Machado A	plunging jet, rocky bed	-	■	■	■	-	-	-	-	-	-	■				-	-	-	-	-	■	-	-	-
	1982	Machado B	plunging jet, rocky bed	-	■	■	■	-	-	-	-	-	-	■				-	-	-	-	-	■	-	-	-
	1985	Mason & Arumugam*	plunging jet	-	■	■	■	■	-	-	-	-	-	■				-	-	-	-	-	-	-	-	-
	1989	Mason	plunging jet	-	■	■	■	■	-	-	-	-	-	-				-	-	-	-	-	■	-	-	-
semi-empirical	1960	Mikhalev	plunging jet	-	■	■	■	■	-	-	-	-	-	■				-	-	-	-	-	-	-	-	-
	1967	Mirtskhulava et al.*	plunging jet, rocky bed	-	-	■	■	■	■	-	-	-	-	-	■	■		-	-	-	-	-	■	-	-	-
	1967	Poreh & Hefez	circ. submerged imp. jet	-	-	■	-	■	-	-	-	-	-	■				-	-	-	-	-	-	-	-	-
	1975	Zvorykin	ski-jump	-	-	■	-	■	-	-	-	-	-	■				-	-	-	-	-	-	-	-	-
	1983	Mih & Kabir	circ. submerged imp. jet	-	-	■	-	■	■	-	-	-	-	■				-	-	-	-	-	-	-	-	-
	1985	Chee & Yuen	plunging jet	-	■	■	■	■	-	-	-	-	-	■				-	-	-	-	-	-	-	-	-
	1985	Spurr*	plunging jet	■	■	-	■	-	■	-	-	-	-	■			■	■	■	■	■	-	-	-	-	-
	1991	Bormann & Julien*	grade-control, plung. jet	-	■	■	-	■	■	■	-	-	-	■				-	-	-	■	-	-	-	-	-
	1993	Stein et al.	plunging jet	■	■	■	-	■	-	-	-	-	-	■				-	-	-	-	-	-	-	-	-
	1994	Fahlbusch	general	-	■	-	■	-	■	-	-	-	-	■			■	-	-	-	-	-	■	-	-	-
	1998	Annandale & al.*	general	-	■	■	-	■	-	-	-	-	-	■			■	■	■	■	■	■	■	-	-	-
	1998	Hoffmans	general	-	■	■	-	■	-	-	-	-	-	■				-	-	-	-	-	-	-	-	-
plunge pool pressure fluctuations	1983	Xu Duo Ming	rectang. impinging jet	-	■	■	■	■	■	-	■	■	-	-				-	-	-	-	-	-	-	-	-
	1985	Cui Guang Tao	rectang. impinging jet	-	■	■	■	■	■	-	■	■	-	-				-	-	-	-	-	-	-	-	-
	1987	Franzetti & Tanda	circular impinging jet	-	■	■	■	■	■	-	■	■	-	-				-	-	-	-	-	-	-	-	-
	1991	Armengou	rectang. falling nappe	-	■	■	■	■	■	-	■	■	-	-				-	-	-	-	-	-	-	-	-
	1991	May & Willoughby	rectangular slot jet	-	■	■	■	■	■	-	■	■	-	-				-	-	-	-	-	■	■	■	■
	1994	Puertas & Dolz	rectang. falling nappe	-	■	■	■	■	■	-	■	■	-	-				-	-	-	-	-	-	-	-	-
	1997	Ervine & al.	circular impinging jet	-	■	■	■	■	■	-	■	■	-	-				-	-	-	-	-	■	■	■	■
pressure difference techniques	1963	Yuditskii	oblique imp. rect. jet	-	-	-	■	-	■	-	-	-	-	-				-	-	-	■	-	-	-	-	-
	1986	Reinius	parallel flow impact	-	■	-	■	-	■	-	-	-	-	■				-	-	■	-	-	-	-	-	-
	1989	Otto	oblique imp. rect. jet	-	-	-	■	-	■	-	-	-	-	■				-	-	■	■	-	-	-	-	-
	1992	Fiorotto & Rinaldo	concrete slab uplift	-	■	■	■	■	-	-	■	■	-	-				-	-	-	■	-	-	-	-	-
	1998	Liu & al.	rock block uplift	-	■	■	■	■	-	-	■	■	-	-				-	-	-	■	-	-	-	-	-
	1999	Liu & al.	vibration. slab uplift	-	■	■	■	■	-	-	■	■	-	-				-	-	-	■	-	-	-	-	-
	2000	Fiorotto & Salandin	anchored slab uplift	-	■	■	■	■	-	-	■	■	-	-				-	-	-	■	-	-	-	-	-
PFL Project	2001		2-phase transient jacking/uplift	■	■	■	■	■	■	■	■	■	■	-	-	■	■	■	■	■	■	■	■	■	■	■

Table 1. Summary of existing scour evaluation methods (Bollaert & Schleiss, 2001c)

The complex scouring process is reduced by the empirical formula to a few parameters. The ultimate total scour depth Y measured from tailwater level (see Fig. 1) is thought to be a function of

- the specific discharge q (discharge per unit width of jet)
- the fall height H
- the tailwater depth h (measured from initial river bed level)
- the characteristic sediment size or rock block diameter d.

Most of the formulas are written in the form

$$Y = t + h = K \cdot \frac{H^y \cdot q^x \cdot h^w}{g^v \cdot d_m^z} \qquad (1)$$

where t is the scour depth below the initial bed level and K a constant.

Mason & Arumugam (1985) applied this form of formula to a large number of scour data, 26 sets from prototypes and 47 from model tests, and found the following best fit exponents and constants for both model and prototype conditions:

$K = (6.42 - 3.10H^{0.10})$
$v = 0.30$
$w = 0.15$
$x = (0.60 - H/300)$
$y = (0.15 - H/200)$
$z = 0.10$

According the data sets the formula is applicable for model fall heights H between 0.325 m and 2.15 m and between 15.82 m and 109 m for prototypes in the case of free overfall jet, ski-jump or orifice spillways. The use of the mean particle size d_m gave better results than the use of the d_{90} particle size. For prototype rock, an equivalent particle size d_m of 0.25 m is recommended in the above formula.

3.3 *Semi-empirical equations*

As already mentioned, semi-empirical methods are combining laboratory and field observations with some physics as:
- initiation of motion of the bed material by shear stress
- energy conservation equations
- geomechanical characteristics
- angle of impingement of the jet
- steady-state two-dimensional jet diffusion theory
- aeration effects

A detailed overview of these semi-empirical equations and methods can be found in Whittaker & Schleiss (1984) and Bollaert (2002). The hydrodynamic process of scour is often derived from the two-dimensional jet diffusion theory. The geomechanical behavior of the rock mass is considered by the shear-stress based initiation of motion concept for non-cohesive granular materials or by an index that defines the resistance of the rock mass against erosion. Both hydrodynamic and geomechanic characteristics are for example combined in Spurr's (1985) and Annandale's (1995) erodibility index methods for rocks and in the momentum conservation equations established by Fahlbush (1994) and Hoffmans (1998) for non-cohesive material.

3.4 *Approaches based on extreme values of fluctuating pressures at the plunge pool bottom*

At the plunge pool bottom fluctuating, dynamic pressures occur due to the direct jet core impact in the case of relatively small water cushion. For high plunge pool depth (higher than 4 to 6 times the thickness of the incoming jet) a turbulent shear flow or developed jet impact according to the two-dimensional jet diffusion theory is created. These two types of jet impact generate completely different pressure patterns as already mentioned.

The dynamic pressures at the plunge pool bottom can penetrate into fissures of the underlying rock mass. The approaches based on the extreme pool bottom pressures assume that the maximum pressures occurring at the water-rock interface are transferred through the joints underneath the rock blocks. These maximum pressures underneath the rock blocks combined with the minimum pressures at the plunge pool bottom create a net uplift pressure Δp on the rock blocks (Figure 4). The ultimate scour is reached when this net pressure difference Δp on the rock block is not able anymore to eject it. Since the maximum and minimum pressures are not occurring at the same moment, the so defined net uplift pressure represents a physical upper limit of dynamic loading conditions and is therefore rather conservative.

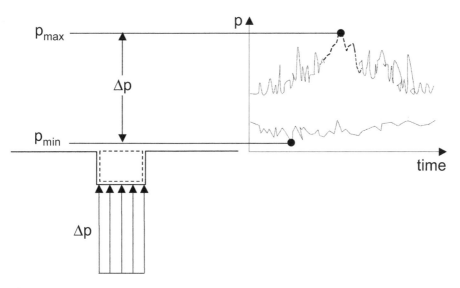

Figure 4. Definition sketch of extreme dynamic pressures at the plunge pool bottom. The maximum and minimum pressure are defined at the center of the block for a long enough time interval (according Bollaert 2002).

Studies on pressure fluctuations in plunge pools have mainly been carried out by Ervine et al. (1997), Xu-Duo-Ming (1983) and Franzetti & Tanda (1984, 1987) for circular jet impingement and by Tao et al. (1985), Lopardo (1988), Armengou (1991), May & Willoughby (1991) and Puertas & Dolz (1994) on rectangular jets. These studies give useful information on bottom pressure fluctuations but do not describe their propagation inside the joints of the underlying rock mass. The simultaneous application of extreme minimum and maximum bottom pressures above and underneath rock blocks can result in net pressure differences of up to 7 times the root-mean-square value or up to 1.5 - 1.75 times the incoming kinetic energy of the jet. Even if this seems to be a conservative design criterion it has to be noted, that violent transient pressure phenomena, which could occur inside the rock joints, are not considered (Bollaert 2002).

3.5 Techniques based on time-mean and instantaneous pressure differences and accounting for rock characteristics

Contrary to the approach presented in the previous chapter, time-averaged or instantaneous pressure differences occurring at or during a certain time at the surface and underneath the rock blocks are considered. This means that the fluctuating pressures have not only to be measured at the plunge pool bottom but also inside the rock joints.

Yuditskii (1963) (reported by Gunko et al., 1965) was probably the first who stated that time averaged and pulsating pressures are responsible for rock block uplift in the scour process. He measured the forces on a single rock block on flat plunge pool bottoms due to the impact of a jet produced by a ski-jump spillway in a scale model. Measurements techniques at that time allowed only to obtain time averaged forces. More recently Otto (1989) quantified time-averaged uplift pressures acting on a rock block for plane jets impinging obliquely. Depending on the relative protrusion of the block and the point of jet impact, important surface suction effects occurred, leading to mean uplift pressures of almost the total incoming kinetic energy. Without considering this suction effect, the maximum uplift pressures were still half of the incoming kinetic energy.

The destructive effects of instantaneous pressure differences entering in tiny rock joints was highlighted the first time by Hartung & Häusler (1973). Kirschke (1974) at first performed an analytical and numerical analysis of water hammer propagation in rock joints. Instantaneous pressure differences based on transient flow assumptions were measured and analyzed for the first time for the case of concrete slab linings of stilling basins (Fiorotto & Rinaldo 1992a, Bellin & Fiorotto (1995), Fiorotto & Salandin 2000). However, the scale of the model tests and the data acquisition rate didn't allow measuring any oscillatory or resonance effects in the joint under the slabs.

Annandale et al. (1998) simulated the erosion of fractured rock by the use of lightweight concrete blocks, placed in a series of two layers on a 45° dip angle. Jet impingement confirmed their theory that the erosion threshold criterion for rock and soil material can be defined by means of a geomechanical index, the so-called Erodibility Index. The erosive power of the water can be related to the erosion resistance of the material.

Experimental and numerical studies focussing on fluctuating net uplift forces on simulated rock blocks were also performed by Liu et al. (1998). A design criterion for rock block uplift was given based on a transient flow model but without considering resonance effects (Liu 1999).

All the mentioned studies on concrete slabs and rock blocks take into account the surface pressure field as a function of space and time. But the pressure field underneath is assumed constant over the surface of the element and equal to the pressure at the entrance of the joints i.e. at the surface. Therefore fully transient flow conditions in the joints such as pressure wave reflections and amplifications are neglected by these extreme pressure techniques.

3.6 *Scour model based on fully transient water pressures in rock joints*

Bollaert (2002) measured for the first time the transient pressures in rock joints due to high-velocity jet impact and reproduced them in a numerical model. New phenomena could be observed and explained as the reflection and superposition of pressure waves, resonance pressures and quasi-instantaneous air release and re-solution due to pressure drops in the joint.

The analysis revealed that the pressure wave velocity is highly influenced by the presence of free air bubbles in the joints. These bubbles can be transported by flow from the plunge pool into the joint but also be released from the water during sudden pressure drop below atmospheric pressure.

In open-end joints instantaneous net uplift pressures of 0.8 to 1.6 times the incoming kinetic energy of the impacting jet have been measured. This is significantly higher than any previous assumptions in literature (see also Chapter 3.5) and underlines that transient pressure effect in rock joints are a key physical process for scour formation.

Based on the experimental results and the numerical simulation, a new model for the evaluation of the ultimate scour depth has been developed, which represents a comprehensive assessment of the two physical processes: hydrodynamic fracturing of closed-end rock joints and dynamic uplift of rock blocks. All relevant processes as the characteristics of the free falling jet (velocity and diameter at issuance, initial jet turbulence intensity), the pressure fluctuations at the plunge pool bottom and the hydrodynamic loading inside rock joints are dealt with and compared with the resistance of the rock against crack propagation.

4 DIFFICULTIES ENCOUNTERED WHEN ESTIMATING SCOUR DEPTH

4.1 *Which is the appropriate formula or theory?*

In the feasibility design stage of a dam project normally the easily applicable empirical and semi-empirical formulae are used to get a first idea of the expected ultimate scour depth occurring downstream of the spillway structures during the lifetime. Most of the formulae have been developed for a specific case such as a ski-jump, a free crest overfall or an orifice spillway, and some others are of general applicability. Therefore a careful selection of the appropriate equations should be made for each project. Nevertheless even after this selection, the results of the remaining formulas show very often a wide scatter. In such a case, the engineering decisions can be based on a statistical analysis of the results by using, for example, the average of all formula or the positive standard deviation in a more conservative way. This analysis is carried out for a certain spillway discharge with the corresponding tailwater level by varying the characteristic rock block size (if considered) according to expected geological conditions.

Sometimes prototype scour measures of an existing dam are available, situated in similar geological conditions than the foreseen new project. Then the formula which predicts best this measured scour depth can be identified and be applied to the new project.

4.2 *How model tests should they be performed and interpreted?*

If free jets impinging on rock underlying a plunge pool have to be modeled in a laboratory, in principal three difficulties arise (Whittaker & Schleiss 1984):
a) appropriate choice of a material that will behave dynamically in the model as fissured rock does in the prototype
b) grain size effects
c) aeration effects

In most models the disintegration process, that means the hydrodynamic fracturing of closed end rock joints and splitting of rock in rock blocks, is assumed to have taken place. Thus only the ejection of the rock blocks into the macroturbulent plunge pool flow and the transport of the material from the scour hole is modeled. Reasonable results may be obtained if fissured rock is modeled by appropriately shaped concrete elements (Martins 1973), but their regular pattern and size is not fully representing a rock mass with several intersecting fracture sets. Nevertheless, when modeling rock crashed gravel should be used, having at least a grain size distribution as the expected rock blocks, instead of round river gravel. In any case, model tests can not simulate the break-up of the rock blocks by the ball milling effect of the turbulent flow in the plunge pool. Therefore in the model a mound is formed which is higher and more stable than in the prototype. As a result the prototype scour depth is underestimated. This can be compensated to some extent by choosing the material carefully including downscaling. Normally good predictive results for scour depth can be obtained by using non-cohesive material, but the extent of the scour may be not correct because steep and near vertical slopes can not be reproduced. Therefore the use of slightly cohesive material by adding cement, clay caulk, paraffin wax etc. to the crushed gravel is proposed (Johnson 1977, Gerodetti 1982, Quintela & Da Cruz 1982).

It is known that for grain sizes smaller than 2 to 5 mm, the ultimate scour depth becomes constant (Veronese 1937, Mirtskhulava et al. 1967, Machado 1982). For an acceptable scale of a comprehensive dam model of 1: 50 to 1:70, the smallest prototype rock blocks, which can be reproduced, are in the range of 0.1 to 0.35 m.

Finally air entrainment cannot be scaled appropriately unless using an unpractical large scale (in the order of 1:10) and has a highly random character, which influences the scour process considerably (see Chapter 2).

4.3 *How to analyze prototype observations properly?*

When analyzing prototype observations on scour depth in order to derive an equation for similar conditions, the following questions have to be answered:
1. What was the duration of the operation of the spillway for different specific discharges (discharge-duration curve)? An example of discharge-duration curve of a spillway is given in Figure 5.
2. Which was the prevailing, specific discharge which formed the scour depth?
3. Was the duration of this specific discharge long enough to create ultimate scour depth?

Since in practice it is very often difficult to answer precisely to these questions, probably significant uncertainties have been introduced into the existing formulas derived from prototype observations. This may also explain the large scatter when predicting scour for other prototypes.

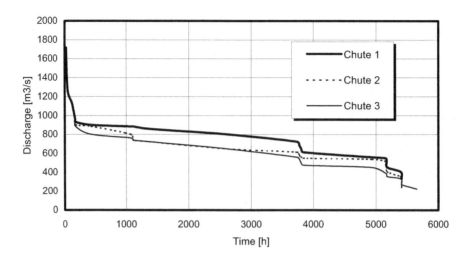

Figure 5. Discharge-duration curve of the chute spillway of Karun I Dam in Iran for the period between March 1980 and July 1988 (width per chute 15 m).

4.4 *Can ultimate scour depth form during operation and what is the scour rate?*

In principle, the scour depth estimated by use of empirical and semi-empirical formula will occur only for a long duration of spillway operation, after steady conditions in the scour hole are achieved. This will happen only after a minimum duration of spillway operation, mainly depending on the quality and jointing of the rock mass. Therefore the specific discharge which will have a sufficiently long duration to form the ultimate scour during the technical life of the dam has to be known. Higher and therefore rare discharges are not able to create ultimate scour.

Since plunge pool scour $t + h$ is known to develop at an exponential rate with time T, the scour rate can estimated with the following relationship (Spurr 1985):

$$(t + h)(T) = (t + h)_{end} (1 - e^{-aT/T_e}) \tag{2}$$

where a is a site-specific constant. The evaluation of T_e (instant at which equilibrium is attained) depends on how rapidly hydro-fracturing and washing out of the material from the scour hole will occur, taking into account the primary and secondary rock characteristics. Primary rock characteristics comprise RQD, joint spacing, uniaxial compressive strength, and angle of jet impact compared to main faults or bedding planes; the secondary characteristics are the hardness and degree of weathering (Spurr 1985). Knowing the depth of a scour which occurred during a certain period of operation and estimating the maximum scour depth by one of the formulae, the site specific constant a/T_e can be determined.

As a rough estimation based on some prototype data, ultimate scour is normally attained only after $T_e = 100$ to 300 hours of spillway operation for a certain discharge considered.

It may be concluded, that the ultimate scour depth for a certain specific discharge occurs only if the duration of this discharge is long enough. Scour for a smaller duration can be estimated with an exponential rate relationship.

4.5 *Which will be the prevailing discharge for scour formation during a flood event?*

A flood event and the corresponding discharge curve of the spillway can be characterized by a hydrograph (Figure 6). For all discharges of the hydrograph with a duration shorter than the instant T_e at which equilibrium is attained, the ultimate scour will not be reached. Knowing the scour rate relationship (Equation 2), the prevailing discharge which will produce maximum scour depth during the flood event can be determined. The scour is estimated successively for

14

discharges q_e ($T = T_e$), q_1 ($T_1 < T_e$), q_2 ($T_2 < T_1 < T_e$), ..., q_{Peak} ($T_{peak} < T_i < T_e$). The discharge, which gives the deepest scour, is the prevailing discharge.

It has to be noted that these considerations are valid only for ungated free surface spillways. For gated free surface spillways the discharge may not be directly related to the reservoir inflow but be prescribed by operation rules. When lowering the reservoir level during floods, outflow discharges are higher than reservoir inflow. This can also be the case for pressurized orifice spillways.

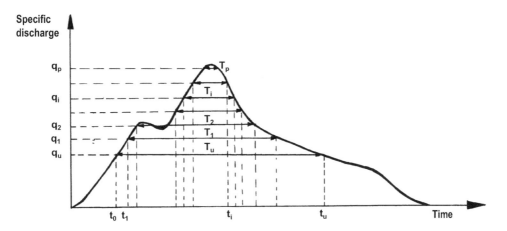

Figure 6. Flood event and the corresponding discharge curve of the spillway showing discharges with a duration shorter than the instant T_e at which equilibrium of scour is attained with the purpose to determine the prevailing discharge.

5 SPILLWAY DESIGN DISCHARGE AND SCOUR EVALUATION

As already mentioned, spillways are designed for the so-called design or project flood, typically a 1000-years flood, and checked for the so-called safety check flood, typically to a flood between a 10'000-years flood and PMF. The question arises for what flood the scour depth has to be evaluated and the constructive scour control measures have to be based on.

As discussed in Chapter 4, the ultimate scour depth will occur only after steady conditions in the scour hole are achieved, which will happen only after a certain duration of spillway operation. Therefore it is very conservative to base the estimation of the scour depth or the design of mitigation measures on low frequency floods (PMF or 1000 years flood). It will be very unlikely, that during the technical life of the dam, these rare floods can produce ultimate scour depth.

Therefore for each flood event with a certain return period, the prevailing discharge and the maximum scour depth has first to be determined according Chapter 4.5. Furthermore it has to be decided, for which flood return period the maximum scour depth during the technical life of the dam shall be estimated. The probability of the occurrence of a flood with a given return period during the useful live of a dam is as follows:

$$r = 1 - (1 - 1/n)^m \qquad (3)$$

where r is the risk or the probability of occurrence, n the return period of the flood (years) and m the useful life of the dam.

In Figure 7 the probability of occurrence of floods for different useful lifetime of the dam is illustrated. It can be seen, that the probability of the occurrence of a 200-years flood during 200 years of operation is 63 %, whereas for a 1000-years flood is only 20 %.

15

It seems reasonable to choose a design discharge with a probability of occurrence of about 50 % during the useful lifetime of a dam for the scour evaluation and the protection measures. Higher design discharges with lower probability of occurrence are too conservative.

It has to be noted, that in the case of gated free surface and orifice spillways, high discharges can be released at any time by opening the gates. Furthermore for low-level outlet spillways the core impact velocity of the jet is nearly independent from discharge.

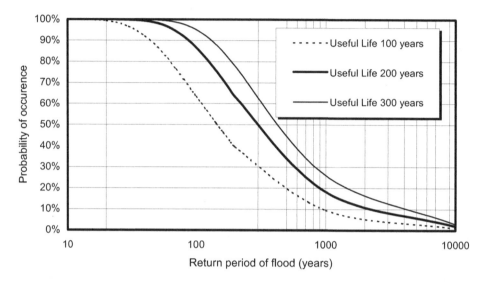

Figure 7. Probability of occurrence of a flood with a given return period for different useful lifetimes of the dam.

6 MEASURES FOR SCOUR CONTROL

6.1 *Overview*

To avoid scour damage, two active options are available (Whittaker & Schleiss 1984):
- avoid scour formation completely
- limit the scour location and extent

Since structures for scour control are rather expensive, only the latter is normally economically feasible (Ramos 1982). The extent of the scour can be influenced by the following measures:
- limitation of the specific spillway discharge
- forced aeration and splitting of jets leaving spillway structures
- increasing tailwater depth by a tailpond dam
- pre-excavation of the plunge pool

The location of the scour depends on the selected type of spillway and its design.

To avoid scour completely structural measures as lined plunge pools are required. Besides the active options, scour damage can be prevented also by passive measures, for example by protecting dam abutments with anchors against instability due to scour formation.

6.2 *Limitation of the specific spillway discharge*

This measure is mainly important in the case of arch dams and free ogee crest spillways. The jet can be guided by an appropriate crest lip design for a given specific discharge at a certain distance from the dam toe. If the dam foundation can be endangered by the scour the discharge per unit length of the ogee crest has to be limited. But by reducing the specific discharge, the available velocity at the crest lip and therefore the travel distance of the jet is also reduced.

In the case of gated ski-jump spillways and low-level outlets, the specific discharge depends on the size of the outlet openings. Since the available velocity at the outlet is high enough to divert the jet far away from the dam and its foundation, the limitation of the specific discharge is normally less important than for free crest spillways.

6.3 *Forced aeration and splitting of jets*

In order to split and aerate the jets leaving flip buckets and crest lips, they are often equipped with baffle blocks, splitters and deflectors. Furthermore high velocity flows in spillway chutes are normally aerated by aeration ramps and slots along the chute. All these measures will increase air entrainment, which will reduce the scouring capacity of the plunging jets. Nevertheless the amount of air entrained is difficult to estimate. Because of scale effects the efficiency of these measures can only be checked qualitatively by hydraulic model tests.

Martins (1973) suggested a reduction of 25 % of the calculated scour depth in the case of high air entrainment and 10 % for intermediate air entrainment. Mason (1989) proposed an empirical expression considering the volumetric air-to-water ratio β. The proposed empirical equation based on spillway models does not depend on the fall height H, since he used a direct relationship between β and H developed by Ervine (1976). Even if physically not very plausible, the empirical formula is accurate for model data and seems to give a reasonable upper bound of scour depth for prototype conditions.

6.4 *Increasing tailwater depth by a tailpond dam*

Another way to control scour from jets is to increase tailwater depth by building a tailpond dam downstream of the jet impact zone. The efficiency of a water cushion is often overestimated (Häusler 1980). For plunge pool depths smaller than 4 to 6 times the jet diameter, core jet impact (see Chapter 2.3) is normally observed at the plunge pool bottom (Bollaert 2002). The jet core is characterized by a constant velocity and is not influenced by the outer two-phase shear layer condition of the impinging jet. The pressures are also constant with low fluctuations, which have significant spectral energy at very high frequencies (up to several 100 Hz). For tailwater depths larger than 4 to 6 times the jet diameter (or thickness), developed jet impact occur at the plunge pool bottom with a different pressure pattern produced by a turbulent two-phase shear layer. Very significant fluctuations are produced with high spectral content at frequencies up to 100 Hz. Maximum fluctuations have been observed for tailwater depths between 5 to 8 times the jet diameter (Bollaert & Schleiss 2001d). Substantial high values persist up to tailwater depth of 10 to 11 times the jet diameter. From this observations may be concluded, that water cushions in the range of 5 to 11 times the jet diameter can generate even more severe dynamic pressures at the plunge pool bottom than smaller tailwater depths. Therefore only water cushions deeper than 11 times the diameter of the jet at impact have a retarding effect on the scour formation.

This tendency was also confirmed by the empirical scour equation of Martins (1973), which gives a maximum scour depth for a certain tailwater depth. Johnson (1967) already found that too small water cushions are even worse than no cushion since the material can be transported more easily out of the plunge pool.

It has to be noted, that increased tailwater level by the help of a tailpond dam may interfere with bottom outlets.

6.5 *Pre-excavation of the plunge pool*

In principal pre-excavation increases the tailwater depth and in view of scour control the same remarks are valid than given in Chapter 6.4.

In general the pre-excavation of the expected scour may be also appropriate, when the eroded and by the river transported material can form dangerous deposits downstream, for example near the outlet of the powerhouse. Such deposits could increase tailwater level and reduce power production. Such problems normally have not to be expected, if the scour is formed about 200 m upstream of the powerhouse outlet.

Pre-excavation of the scour hole is also often considered when instabilities of the valley slopes have to be feared. In such cases the excavation has to be stabilized by anchors and other measures. As an alternative, even under such conditions, pre-excavation could be omitted, if the valley slopes are stabilized by appropriate measures in such a way, that the slopes be stable even after ultimate scour formation.

The selection of the design discharge for an excavated geometry of the plunge pool has to be based on considerations similar to the ones discussed in Chapter 5. In general it is economically not interesting to excavate deeper than the scour depth which would be formed by a 50 to 100 years flood. If instabilities of the valley slopes, in case of a deeper scour due to higher spillway discharges, have to be feared, rock anchors and pre-stressed tendons can be used, as already mentioned (Figure 8).

The pre-excavated plunge pool geometry has to be based on the expected natural scour geometry. Several authors proposed empirical formula for the estimation of the length (Martins 1973, Kotoulas 1967) and the width (Martins 1973) of the scour hole. Amaniam & Urroz (1993) performed a number of tests on a model scale flip bucket spillway with a gravel bed plunge pool in order to develop equations to describe the geometry of the scour hole created by the jet impinging into the plunge pool. They observed that the performance of the pre-excavated scour holes are best if those are close to the self-excavated hole for the same flow parameters.

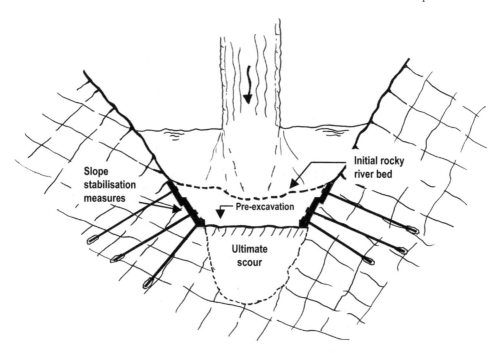

Figure 8. Pre-excavation of the plunge pool for a scour depth, which would be formed by a 50 to 100 years flood and slope stabilization measures in case of ultimate scour formation.

6.6 *Concrete lined plunge pools*

If absolutely no scour formation in the rock downstream of a dam can be accepted, the plunge pool has to be reinforced and tightened by a concrete lining. Since the thickness of the lining is limited by construction and economical reasons, normally high tension or pre-stressed rock anchors are required to ensure the lining stability in view of the high dynamic loading. Furthermore the surface of the lining has to be protected with reinforced (wire mesh, steel fibers) and high tensile concrete having also high resistance against abrasion. Construction joints have to be sealed with efficient water stops. In addition the stability of the lining against static uplift pressure during dewatering of the plunge pool has to be guaranteed by a drainage system. This can also limit dynamic uplift pressures in case of limited cracking of the lining.

The design of plunge pools linings has to be based on the following sequence of events (Fiorotto & Rinaldo 1992, Fiorotto & Salandin 2000):
- Pulsating pressures can damage the joint seals between slabs (construction joints).
- Through these joint seals, extreme pressure values may propagate from the upper to the lower surface of the slabs.
- Instantaneous pressure differentials between the upper and lower surfaces of the slabs can reach high values.
- The resultant force stemming from the pressure differential may exceed the weight of the slab and the anchor's resistance.

Furthermore the propagation of dynamic pressures through fissures in the lining reveals the presence of water hammer effects, which can amplify the pulsating uplift pressures underneath the concrete slabs (Bollaert 2002).

Since cracking of the plunge pool lining can not be excluded, the assumptions of an absolutely tight lining and neglecting dynamic uplift are on the very unsafe side. If the high dynamic pressures can propagate through a small, local fissure from the upper to the lower surface of a concrete slab, dynamic pressures from underneath will lift the slab locally, which finally results in a progressive failure of the whole plunge pool lining. Thus the concept of a tight plunge pool lining is as risky as a chain concept: the system's resistance is given by the weakest link, i.e. the local permeability of the concrete slab.

Furthermore fluctuating pressures due to the plunging jets into the plunge pool are high compared to the proper weight of the slab and thus will result in vibration of the concrete slab and consequently in the development of cracks in the concrete and in possible fatigue failure of the anchors.

Cracking of a concrete slab before operation can not be ruled out even using sophisticated construction joints and waterstops, because of temperature effects when filling the plunge pool with water and of deformation of the underground. Therefore the design criteria of the plunge pool liner have to take into account the following:
- the load case of dynamic uplift pressures during operation
- reinforcement in the concrete slab has to be designed for crack width limitation for possible dynamic vibration modes (depending on anchor pattern and stiffness).
- the grouted and pre-stressed rock anchors have to be designed for fatigue.

The drainage system under the plunge pool lining is of very high importance since it increases considerably safety against dynamic uplift pressures. Nevertheless, since limited cracking of the lining can not be excluded, as already mentioned, pressure waves can be transferred through the cracks into the drainage system. The response of the drainage system in view of these dynamic pressures with a wide range of frequencies at the entrance of a possible crack has to be controlled by a transient analysis, in order to be sure that no amplification of the pressures in the drainage system occurs.

7 CONCLUSIONS

Although the physical understanding of the scour process has considerably improved during the last 10 to 20 years, the scour evaluation in space and time still remains a challenge for dam designers. Scour models are now available which take into account the pressure fluctuations in the plunge pool and the propagation of transient water pressures into the joints of the underlying rock mass. Nevertheless, for the time being only simple plunge pool and rock joint geometries can be considered. Further research is therefore still needed.

Despite the development of complex theoretical models, which consider all the physical processes, there will still remain open questions, which have to be answered on the bases of engineering judgment. Above all, rock parameters as number, spacing, direction and persistency of fracture sets, in-situ stresses in the rock mass, fracture toughness, unconfined compressive strength a.s.o. have to be estimated or determined with reasonable costs all over the space of the expected scour hole. These parameters will also highly influence evolution of scour with time together with the hydrological conditions during the considered lifetime of the dam, which involves further uncertainties.

In order to check and calibrate complex scour models, more detailed prototype data on scour evolution with fully documented discharge records are needed. In principal, these observations are essential for a continuous safety assessment of a dam and allow predicting the scour evolution.

REFERENCES

Amanian, N. & Urroz, G.E. 1993. Design of pre-excavated scour hole below flip bucket spillways. Proceedings of the ASCE International Symposium on Hydraulics, San Francisco.
Annandale, G. W. 1995. Erodibility, Journal of Hydraulic Research, IAHR, Vol. 33, N°4, pp. 471-494.
Annandale, G.W.; Wittler, R.J.; Ruff, J.; Lewis, T.M. 1998. Prototype validation of erodibility index for scour in fractured rock media, Proceedings of the 1998 International Water Resources Engineering Conference, Memphis, Tennessee, United States.
Armengou, J. 1991. Disipacion de energia hidraulica a pie de presa en presas boveda, PhD thesis, Universitat Politechnica de Catalunya, Barcelona.
Ballio, F.; Franzetti, S.; Tanda, M.G. 1994. Pressure fluctuations induced by turbulent circular jets impinging on a flat plate, Excerpta, Vol. 7.
Bellin, A. & Fiorotto, V. 1995. Direct dynamic force measurement on slabs in spillway stilling basins, Journal of Hydraulic Engineering, ASCE, Vol. 121, N° HY 10, pp. 686-693.
Beltaos, S. & Rajaratnam, N. 1973). Plane turbulent impinging jets, Journal of Hydraulic Research, IAHR, Vol. 11, N° 1, pp. 29-59.
Bin, A.K. 1984. Air entrainment by plunging liquid jets, IAHR Symposium on Scale Effects in Modeling Hydraulic Structures, Esslingen.
Bollaert, E. 2001. Spectral density modulation of plunge pool bottom pressures inside rock fissures, Proceedings of the XXIXth IAHR Congress, Student Paper Competition, Beijing.
Bollaert, E. 2002. Transient water pressures in joints and formation of rock scour due to high-velocity jet impact. Communication N° 13 of the Laboratory of Hydraulic Constructions (LCH), EPFL, Lausanne.
Bollaert, E. & Schleiss, A. 2001a. Air bubble effects on transient water pressures in rock fissures due to high velocity jet impact, Proceedings of the XXIXth IAHR Congress, Beijing.
Bollaert, E. & Schleiss, A. 2001b. A new approach for better assessment of rock scouring due to high velocity jets at dam spillways, Proceedings of the 5th ICOLD European Symposium, Geiranger, Norway.
Bollaert, E. & Schleiss, A. 2001c. Scour of rock due to high velocity plunging jets. Part I: a state-of-the-art review, submitted to the Journal of Hydraulic Research, IAHR, Delft, The Netherlands.
Bollaert, E. & Schleiss, A. 2001d. Scour of rock due to high velocity plunging jets. Part II: Experimental results of dynamic pressures at pool bottoms and in one-and two-dimensional closed-end rock joints, submitted to the Journal of Hydraulic Research, IAHR, Delft, The Netherlands.
Eggenberger, W. & Müller, R. 1944. Experimentelle und theoretische Untersuchungen über das Kolkproblem. Mitteilungen Nr. 5 der VAW, ETH Zürich.
Ervine, D.A. 1976. The entrainment of air in water, Water Power and Dam Construction, 28(12), pp. 27-30.
Ervine, D.A. 1998. Air entrainment in hydraulic structures: a review, Proceedings of the Institution of Civil Engineers Wat., Marit. & Energy, Vol. 130, pp. 142-153.
Ervine, D.A. & Falvey, H.R. 1987. Behavior of turbulent jets in the atmosphere and in plunge pools, Proceedings of the Institution of Civil Engineers, Part 2, Vol. 83, pp. 295-314.
Ervine, D.A. & Falvey, H.R. 1988. Aeration in jets and high velocity flows, Conference Proceedings, Model-Proto Correlation of Hydraulic Structures, P. Burgi, 1988, pp. 22-55.
Ervine, D.A.; Falvey, H.R.; Withers, W. 1997. Pressure fluctuations on plunge pool floors, Journal of Hydraulic Research, IAHR, Vol. 35, N°2.
Ervine, D.A.; McKeogh, E.; Elsawy, E.M. 1980. Effect of turbulence intensity on the rate of air entrainment by plunging water jets, Proceedings of the Inst. Civ. Eng., Part 2, pp. 425-445.
Fahlbusch, F. E. (1994). Scour in rock riverbeds downstream of large dams, Hydropower & Dams, pp. 30-32.
Fiorotto, V. & Rinaldo, A. 1992. Fluctuating uplift and lining design in spillway stilling basins, Journal of Hydraulic Engineering, ASCE, Vol. 118, HY4.
Fiorotto, V. & Salandin, P. 2000. Design of anchored slabs in spillway stilling basins, Journal of Hydraulic Engineering, ASCE, Vol. 126, N° 7, pp. 502-512.
Franzetti, S. & Tanda, M.G. 1984. Getti deviati a simmetria assiale, Report of Istituto di Idraulica e Costruzioni Idrauliche, Politecnico di Milano.

Franzetti, S. & Tanda, M.G. 1987. Analysis of turbulent pressure fluctuation caused by a circular imping-ing jet, International Symposium on New Technology in Model Testing in Hydraulic Research, India, pp. 85-91.

Gerodetti, M. 1982. Auskolkung eines felsigen Flussbettes (Modellversuche mit bindigen Materialen zur Simulation des Felsens). Arbeitsheft N° 5, VAW, ETHZ, Zürich.

Gunko, F.G.; Burkov, A.F.; Isachenko, N.B.; Rubinstein, G.L. ; Soloviova, A.G. ; Yuditskii, G.A. 1965. Research on the Hydraulic Regime and Local Scour of River Bed Below Spillways of High-Head Dams, 11th Congress of the I.A.H.R., Leningrad.

Hartung, F. & Häusler, E. 1973. Scours, stilling basins and downstream protection under free overfall jets at dams, Proceedings of the 11th Congress on Large Dams, Madrid, pp.39-56.

Häusler, E. 1980. Zum Kolkproblem bei Hochwasser-Entlastungsanlagen an Talsperren mit freiem Über-fall. Wasserwirtschaft 3.

Hoffmans, G.J.C.M. 1998. Jet scour in equilibrium phase, Journal of Hydraulic Engineering, ASCE, Vol. 124, N°4, pp. 430-437.

Johnson, G. 1967. The effect of entrained air in the scouring capacity of water jets, Proceedings of the 12th Congress of the I.A.H.R., Vol. 3, Fort Collins.

Johnson, G. 1977. Use of a weakly cohesive material for scale model scour tests in flood spillway design, Proceedings of the 17th Congress of the I.A.H.R., Vol. 4, Baden-Baden.

Kamenev, I.A., 1966. Alcance de jactos livres provenientes de descarregadores. (Trans. N° 487 L.N.E.C.) Gidrotekhnicheskoe Stroitel'stvo N° 3.

Kawakami, K. 1973. A study on the computation of horizontal distance of jet issued from ski-jump spill-way. Trans. of the Japanese Society of Civil Engineers, Vol. 5.

Kirschke, D. 1974. Druckstossvorgänge in wassergefüllten Felsklüften, Veröffentlichungen des Inst. Für Boden und Felsmechanik, Univ. Karlsruhe, Heft 61.

Kotoulas, D. 1967. Das Kolkproblem unter besonderer Berücksichtigung der Faktoren "Zeit" und "Geschiebemischung" im Rahmen der Wildbachverbauung. Schweizerische Anstalt für das Forstliche Versuchswesen, Vol. 43, Heft 1.

Liu, P.Q. 1999. Mechanism of energy dissipation and hydraulic design for plunge pools downstream of large dams, China Institute of Water Resources and Hydropower Research, Beijing, China.

Liu, P.Q.; Dong, J.R.; Yu, C. 1998. Experimental investigation of fluctuating uplift on rock blocks at the bottom of the scour pool downstream of Three-Gorges spillway, Journal of Hydraulic Research, IAHR, Vol. 36, N°1, pp. 55-68.

Lopardo, R.A. 1988. Stilling basin pressure fluctuations, Conference Proceedings, Model-Prototype Cor-relation of Hydraulic Structures, P. Burgi, pp. 56 – 73.

Machado, L.I. 1982. O sistema de dissipação de energia proposto para a Barragem de Xingo, Transactions of the International Symposium on the Layout of Dams in Narrow Gorges, ICOLD, Brazil.

Martins, R. 1973. Contribution to the knowledge on the scour action of free jets on rocky river beds, Pro-ceedings of the 11th Congress on Large Dams, Madrid, pp. 799-814.

Martins, R. 1977. Cinemática do jacto livre no âmbito das estruturas hidráulicas. Memória N° 486, L.N.E.C., Lisboa.

Mason, P. J. 1989. Effects of air entrainment on plunge pool scour. Journal of Hydraulic Engineering, ASCE, Vol. 115, N° 3, pp. 385-399.

Mason, P. J. & Arumugam, K. 1985. Free jet scour below dams and flip buckets, Journal of Hydraulic Engineering, ASCE, Vol. 111, N° 2, pp. 220-235.

May, R.W.P. & Willoughby, I.R. 1991. Impact pressures in plunge pool basins due to vertical falling jets, Report SR 242, HR Wallingford, UK.

Mirtskhulava, T.E.; Dolidze, I.V.; Magomeda, A.V. 1967. Mechanism and computation of local and gen-eral scour in non cohesive, cohesive soils and rock beds, Proceedings of the 12th IAHR Congress, Vol. 3, Fort Collins, pp. 169-176.

Otto, B. 1989. Scour potential of highly stressed sheet-jointed rocks under obliquely impinging plane jets, PhD thesis, James Cook University of North Queensland, Townsville.

Puertas, J. & Dolz, J. 1994. Criterios hidraulicos para el diseno de cuencos de disipacion de energia en presas boveda con vertido libre por coronacion, PhD thesis, Politechnical University of Catalunya, Bar-celona, Spain.

Quintela, A.C. & Da Cruz, A.A. 1982. Cabora-Bassa dam spillway, conception, hydraulic model studies and prototype behaviour, Transactions of the International Symposium on the Layout of Dams in Nar-row Gorges, ICOLD, Brazil.

Ramos, C.M. 1982. Energy dissipation on free jet spillways. Bases for its study in hydraulic models, Transactions of the International Symposium on the Layout of Dams in Narrow Gorges, ICOLD, Rio de Janeiro, Brazil, Vol. 1, pp. 263-268.

Spurr, K. J. W. 1985. Energy approach to estimating scour downstream of a large dam, Water Power & Dam Construction, Vol. 37, N°11, pp. 81-89.

Tao, C.G.; JiYong, L.; Xingrong, L. 1985. Translation from Chinese by de Campos, J.A.P.. Efeito do impacto, no leito do rio, da lâmina descarregada sobre uma barragem-abóbada, Laboratório Nacional de Engenharia Civil, Lisboa.

Toso, J. & Bowers, E.C. (1988). Extreme pressures in hydraulic jump stilling basin. Journal of Hydraulic Engineering, ASCE, Vol. 114, N° HY8, pp. 829-843.

Taraimovich, I.I. 1980. Calculation of local scour in rock foundations by high velocity flows, Hydrotechnical Construction N°8.

U.S.B.R. 1978. Hydraulic design of stilling basins and energy dissipators. Water Resources Technical Publication. Engineering Monograph N° 25, 4th Printing.

Van de Sande, E. & Smith, J.M. 1973. Surface entrainment of air by high velocity water jets, Chem. Engrg. Sci., 28, pp. 1161-1168.

Veronese, A. 1937. Erosion of a bed downstream from an outlet, Colorado A & M College, Fort Collins, United States.

Whittaker, J. & Schleiss, A. 1984. Scour related to energy dissipators for high head structures, Zürich.

Xu-Duo-Ming 1983. Pressão no fundo de um canal devido ao choque de um jacto plano, e suas características de fluctuação, Translation from Chinese by J.A. Pinto de Campos, Lisboa.

Yuditski, G.A. 1963. Actual pressure on the channel bottom below ski-jump spillways, Izvestiya Vsesoyuznogo Nauchno – Issledovatel – Skogo Instituta Gidrotekhiki, Vol. 67, pp. 231-240.

Zvorykin, K.A., Kouznetsov, N.V., Akhmedov, T.K. 1975. Scour from rock bed by a jet spilling from a deflecting bucket of an overflow dam. 16th Congress of the IAHR, Vol.2, São Paulo.

Case studies and prototype observations of scour

Jiroft Dam, Iran
(Courtesy of Stucky Ingénieurs-conseils SA)

Rock Scour due to falling High-velocity Jets - Schleiss & Bollaert (eds)
© 2002 Swets & Zeitlinger, Lisse, ISBN 90 5809 518 5

Review of plunge pool rock scour downstream of Srisailam Dam

P.J. Mason
Binnie Black & Veatch, Redhill, Surrey, UK

ABSTRACT: The writer has used his own formulae, itself a derived improvement on 31 formulae by others, to estimate scour depth at Srisailam dam. The formula proved to be a reasonable fit to the measured scour depths at Srisailam and the associated unit flows. The formula was calibrated using key flood events to derive an equivalent bed particle size of 0.65 m as characterising the scour process at Srisailam. On this basis the probable deepest scour elevation at Srisailam, for the design flood, was determined. There is a certain asymmetry to the development of the plunge pool with the right hand side generally deeper than the left. Over the central section of the pool there appears to be a dominant upstream slope to the plunge pool of about 14°. This slope, projected from a central section of pool deepened to the maximum predicted elevation, would not seem to undercut the main apron.

1 INTRODUCTION

In December 2000 the present writer, on behalf of Binnie Black & Veatch (BBV), was retained by the Irrigation & CAD (I&CAD) Department of the Government of Andhra Pradesh, India, to advise on the bed-rock erosion, which has occurred in the spillway plunge pool, immediately downstream of the Srisailam dam. This paper presents some of the results from those studies. The information is presented principally in metric units, however both metric and imperial units are in use at Srisailam, therefore some data is also quoted in imperial units.

2 BACKGROUND AND DATA

Srisailam dam is sited across the river Krishna, in Kurnool District of Andhra Pradesh. It is about 110 km upstream from Nagarjunasagar dam and about 200 km from Hyderabad. The project consists of a 143.25m high masonry dam and a right bank power station equipped with 7 units of 110 MW each. The power station is located about 457m downstream of the dam on the right flank. The reservoir formed by the dam has a full storage level of 269.75m, a capacity of 8,720 million m^3 and a surface area of about 617 sq.km. A left bank power station, with an installed capacity of 900 MW, is under construction.

The spillway is located in the deep river portion with non-overflow blocks on either flank. It has an overall operating crest length of 266.4m at El. 252.984m. It comprises 12 spans of 18.30m (60 ft) clear width, equipped with radial crest gates of 18.30m x 21.49m. It is designed to discharge a maximum outflow discharge of 38,370 cumecs at maximum reservoir level of El. 271.88m. A ski-jump bucket with invert at El. 188.976m has been provided as an energy dissipator at the downstream toe of the dam. Two sluices provided in the right flank non-overflow portion are designed to pass a maximum discharge of 1,500 cumecs.

The spillway has discharged floods since 1984 with major floods of 16,008 cumecs in 1994 and 24,471 cumecs in 1998. Discharges throughout the life of the scheme have been focussed on the central gates. In fact 70% of all flows have been passed through the central 4 gates.

3 SCOUR FORMULAE AND GENERAL EXPERIENCES

Ski-jump spillways project free trajectory jets downstream of the dam or chute to a point where they will erode a scour hole, or plunge pool in the river bed-rock. This remains probably the most cost effective and safest way of dissipating hydraulic energy for high head dam spillways. The forces involved can be massive. It can be shown that at the peak of the 1998 Srisailam dam flood, the power of the jet entering the plunge pool was approximately 20,000 MW. Not surprising, erosion took place.

A brief mention should be made of the pattern of currents in a typical plunge pool, as these are often misunderstood. They are illustrated in Figure 1.

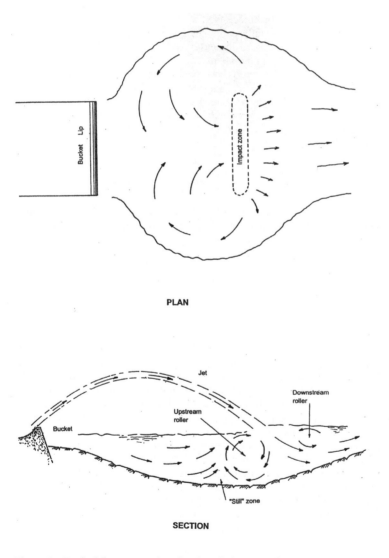

Figure 1. Typical flow pattern in a developed plunge pool.

The jet entering the pool is deflected primarily upstream, however some deflection also occurs downstream. The jet also induces "roller currents" both immediately downstream and upstream. The tendency is to project flow downstream with tailwater levels downstream of the jet higher than those upstream. This in turn induces strong lateral side currents either side of the jet from downstream to upstream. These feed back into the jet from downstream. This induces flow feeding from downstream and also effectively "traps" the roller immediately downstream of the jet, limiting its upstream extent. In fact, where the downstream returning current meets the downstream roller there is a "dead" or "still" area where the respective flows cancel each other out. This was clearly demonstrated by underwater video photography at Srisailam. This showed a zone where the bed was strewn with small loose pebbles, which, anywhere else in the plunge pool, would have been washed away. This point was about 700 ft downstream of the dam reference line whereas jet impact on the bed is at around 850 to 950 ft downstream of the same reference line.

4 SCOUR DEPTH

Previous work by the previous work by the present writer (Mason & Arumugam 1995a) showed that scour depth is broadly proportional to unit flow to the power 0.6. This led to the development of the writer's own formula:

$$D = 3.27 \left[\frac{q^{0.60} \quad H^{0.05} \quad h^{0.15}}{g^{0.30} \quad d^{0.10}} \right] \tag{1}$$

where
D = Scour depth below tailwater level (m)
q = Unit flow (cumecs/m)
H = Head drop reservoir level to tailwater level (m)
h = Tailwater depth (m)
g = Gravitational constant (9.81 m/s/s)
d = Characteristic size of bed material (m)

The scour elevations reached each year at Srisailam dam are given in Figure 2. To compare the formula above with recorded data for Srisailam, actual annual scour depths, and depths predicted using Equation (1), were plotted against $q^{0.6}$. The results are shown on Figure 3.

As a further refinement the history of scour was reviewed on a year-by-year basis using Figure 2. It can be seen from Figure 2 that "defining events" occurred in years 1986, 1988 and 1998 in that scour deepened noticeably as a result of the unit flows reached. Some deepening may have also occurred in 1990 and 1993, though this is less clear as the plunge pool remained at broadly the same depth from 1988 to 1993.

Figure 3 features the scour depths for years 1986, 1988 and 1998 shown as black discs. Years 1990 and 1993 are shown as open circles. It can be seen that the writer's formulae can be adapted to give a very good line of fit to the characteristic years. This was achieved by using a characteristic particle size to 0.65m. This is considered to represent the defining curve for future scour increases should larger floods occur. From this is can be shown that the predicted scour depth for the design flood of 1,355,000 cusecs (38,369 cumecs) would correspond to a deepest scour elevation 387 ft.

It should be noted that the size of 0.65m is simply a mechanism for obtaining a line of best fit through the characteristic years. The nature of the bed material is likely to be more complex than can be represented by a single parameter. Nevertheless it is considered adequate for present purposes. It should also be noted that various rock cores were inspected at site, including one from the downstream bed. Although many of the formations at the site appear massive in nature, hardly any pieces of core were longer than 0.3m. This suggests a high degree of internal fracturing or weakness, which will yield on disturbance, either by drilling or under jet action. On the left side of the plunge pool the rocks above water level also appears blocky and massive. However, noticeable degradation and splitting of these rocks has occurred at the immediate area of jet impact, again highlighting the above point. It is therefore considered that representing the bed by a characteristic particle size of 0.65m may be both convenient and realistic.

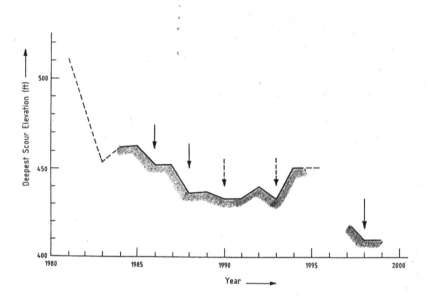

Figure 2. Deepest scour elevation plotted against time.

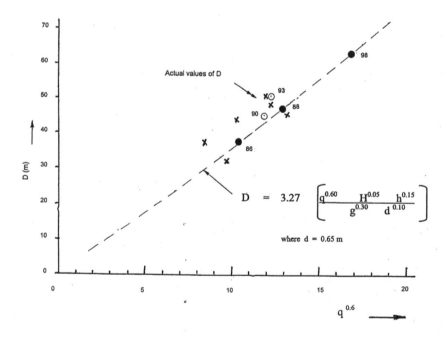

$$D = 3.27 \left[\frac{q^{0.60}}{g^{0.30}} \frac{H^{0.05}}{d^{0.10}} h^{0.15} \right]$$

where d = 0.65 m

Figure 3. Comparison between actual scour depths and scour depths calculated using the author's formula and assuming an equivalent bed material size of 0.65 m.

28

5 POOL PROFILES

Extensive bathymetric surveying of the pool took place in 2000 by the Indian National Oceano-graphic Institute (NOI) based in Goa. Based on this, Figure 4 represents longitudinal sections on the centrelines of gates 4 and 9 and between gates 6 and 7. This represents the most developed, central, zone of the pool area. In fact the pool has developed more deeply on the right side, possibly as a result of original construction flows also being diverted on that side, and because the flow exits on the right side.

A few approaches have been developed for estimating the upstream slope of plunge pools, but the only ones with which the present writer has had any success are those by Blaisdell and those by Taramovich. However, Blaisdell's work can really only be adapted for narrow spillway chutes. The author Taramovich (see Mason 1993) established curves for estimating the up-stream slope angle of scour holes based on jet impact angle. The present writer used these curves with success at another (Peace Canyon) dam, which featured a shale bedrock with clay seams. At Srisailam they would suggest a probable upstream slope angle for the scour hole of about 24° to the horizontal.

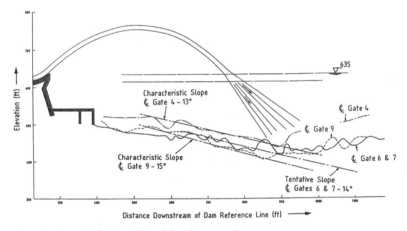

Figure 4. Longitudinal profiles of the plunge pool.

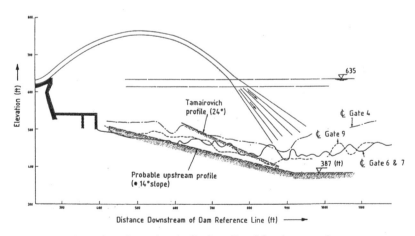

Figure 5. Estimated maximum longitudinal profile of the plunge pool.

Figure 4 also puts lines of best-fit to the upstream slopes of the scour hole, between gates 4 and 9. The slopes range from 13 to 15° with an average of 14°. This is clearly much flatter than the Taramovich slope of 24°. It is not clear whether this is a result of the earlier construction floods, a characteristic of the bed material, or the result of asymmetrical pool development exacerbating return current development. It could be a function of all three.

Figure 5 takes the design flood scoured bed level of 387ft and projects it upstream at 14°. The projection point also allows for some upstream dispersion of the jet. It can be seen that the slope reaches the dam apron at approximately level 515ft. This suggests that the hydraulic action of the plunge pool, including return currents, should not be such as to inherently undermine the apron nor, by inference, the downstream toe of the dam.

The extent of lateral scour is more difficult to estimate. It is also difficult to interpret from what has happened so far at Srisailam where outer gates 1 and 12 are not used and over 70% of all flows have been concentrated over the central third of the spillway. The central area of the pool also features high, steep slopes above water on the left flank, but rubble below water. The right side is flatter above water level but features escarpments and cave features below water. Typically pools deepen initially, under the immediate impact of the jet and subsequently widen as the steep side faces then relax. Downstream of narrow chutes, scour holes often appear almost circular in plan with widths of 2 to 3 times the width of the bucket.

6 CONCLUSIONS

The writer has used his own formulae, itself a derived improvement on 31 other formulae by others, to estimate scour depth at Srisailam dam. The formula proved to be a reasonable fit to the measured scour depths at Srisailam and the associated unit flows. The formula was then calibrated using key flood events to derive an equivalent bed particle size of 0.65m as characterising the scour process at Srisailam. On this basis the probable deepest scour elevation at Srisailam, for the design flood, will be approximately 387 ft.

Flood flows have been passed preferentially over the central third of the spillway. There is also a certain asymmetry to the development of the plunge pool with the right hand side generally deeper than the left. It is suspected that this is causing the development of asymmetrical return currents within the pool.

Over the central section of the pool it can be shown that there is a dominant upstream slope to the plunge pool of about 14°. This slope, projected from a central section of pool deepened during the design flood to 387 ft, would not seem to undercut the main apron.

Due to the lack of use of the end gates, the pool has not spread laterally much beyond the confines of the river. In a fully developed pool, such lateral spreading is not uncommon. Based on experience elsewhere, a lateral spreading at river level into either bank would not be untoward in the longer term, however, it will be very dependant on the nature of the rock.

7 ADDITIONAL COMMENTS

Aspects not covered in the previous pages include the probable accuracy of the estimates made. The discussion below therefore attempts to address these.

The depth of scour has been estimated using a formula, which is itself an improvement on 31 others and has also been calibrated against events on site. Factors causing further deepening could theoretically include further degradation of the bedrock to a smaller particle size and/or more prolonged, high intensity flows. Given that the design flood is, in theory, a one-off event, it would seem appropriate to calibrate it against the principal floods already assessed. This would imply that the 387 figure is realistic. It should also be noted that the deepest levels assumed for calibration are localised spot levels.

The upstream slope is more difficult to confirm as there is little established guidance from elsewhere. However, given that the adopted slope of 14° is so much flatter than the "Taraimovich" slope of 24°, it would seem sufficiently conservative. Note also that the slope has been assumed to start from some point upstream of the direct jet trajectory, which is itself a conservative assumption.

The lateral spread of the jet is the most difficult to assess. It is, however, advisable to arrange for such enlargement to happen in a controlled way, when gates can still be closed, if required, rather than during a large flood when such detailed control becomes difficult. It is quite possible that left and right flanks will erode to different extents and in slightly different forms. Modified gate operational patterns could be used in future to facilitate a controlled lateral development of the pool.

8 ACKNOWLEDGEMENTS

The writer would like to express his sincere appreciation of all those that he encountered from the Government of Andhra Pradesh, the Central Water Commission and the Dam Safety Review Panel in India. They are not listed by name as they are too numerous to mention individually. The writer was impressed by their courtesy and dedication and by their painstaking collection of the data that made this work possible. It is always a pleasure to work with such professionals.

REFERENCES

Mason, P.J. 1982. "The Choice of Hydraulic Energy Dissipator for Dam Outlet Works based on a Survey of Prototype Usage", Proc. of the Inst. of Civ. Engrs., Part 1, (72), May 1982.

Mason, P.J. 1983. Scour Downstream of Energy Dissipating Spillways, The City University, London, April 1983. Prepared in partial fulfilment of the Degree of Master of Philosophy.

Mason, P.J. 1984. "The Erosion of Plunge Pools Downstream of Dams due to the action of Free Trajectory Jets", Proc. of the Inst. of Civ. Engrs., Part 1, (76), May 1984.

Mason, P.J. and Arumugam, K. 1985a. "Free Jet Scour Below Dams and Flip Buckets", Am. Soc. of Civ. Engrs., Jrnl. of Hydr. Div., 1 1 1(2), Feb 1985.

Mason, P.J. and Arumugam, K. 1985b. "A Review of 20 years of Scour Development at Kariba Dam", 2nd It. Conf. on the Hydr. of Floods and Flood Control, paper A5, BHRA, Cambridge, England, 24-26 Sept 1985.

Mason, P.J. 1987. "Scour under Air Entrained Jets Below Dams and Flip Buckets", The City University, London, April 1987. Prepared in partial fulfilment of the degree of Doctor of Philosophy.

Mason, P.J. 1989. "Effects of Air Entrainment on Plunge Pool Scour, Am. Soc. of Civ. Engrs., Jrnl of Hydr. Div., 115(3), March 1989.

Mason, P.J. 1993. "Practical Guidelines for the Design of Flip Buckets and Plunge Pools", Int. Water Power & Dam Constr., 45 (911 0), Sept/Oct 1993.

Scour hole geometry for Fort Peck spillway

A.F. Babb & D. Burkholder
Northwest Hydraulic Consultants, Vancouver, B. C., Canada

R.A. Hokenson
R. W. Beck, Inc., Seattle, WA, U.S.A.

ABSTRACT: A procedure was developed to predict depth and extent of plunge pool scour downstream from Fort Peck Dam. The method used a standard maximum-depth equation calibrated with low-flow prototype measurements and high-flow model tests conducted in 1935 to predict scour for flows ranging up to the probable maximum flood.

1 SPILLWAY GEOMETRY AND FLOW RATES

Fort Peck Dam is owned and operated by the U. S. Army Corps of Engineers (COE). This paper describes a procedure developed to predict the depth and extent of plunge pool scour produced by flows passing over the Fort Peck spillway that range up to the probable maximum flood.

Fort Peck Spillway (Figure 1) consists of: (1) a gated overflow weir with a net crest length of 195 m, (2) a 1,524 m long concrete-lined trapezoidal-shaped chute with 1:1 sidewalls converging to a bottom width of 36.6 m with a terminal slope of 5.23 %, and (3) a terminal structure consisting of a 21.3 m deep cutoff wall with 80 m long wingwalls set back at a 45^0 angle. The overall vertical drop from the spillway crest to the downstream end of the spillway is 65.2 m. The spillway has no energy dissipation structure. Foundation material for the spillway is highly erodible Bearpaw shale.

Spillway flows are predicted to reach a maximum of 7,500 m^3/s during passage of the probable maximum flood. The prototype spillway has operated infrequently since its construction in the early 1940's, but during the period from 1946 to 1953, during filling of the downstream Missouri River dams, the spillway was operated with flows ranging from about 280 to 765 m^3/s. In August and September of 1946, 307 Mm^3 were released with a maximum flow of 765 m^3/s for a period of less than one day. The spillway operated for 90 days for each of the three years 1947 - 1949 and from 60 to 90 days for each of the years 1950 to 1953. The spillway was also operated with flows up to 566 m^3/s during 1975, 1976, and 1979 for durations from 30 to 60 days per year. A summary of flows with maximum scour depth and eroded volume is shown in Table 1.

2 PROTOTYPE SCOUR

Prototype scour measured for the years from 1946 to 1953 documented scour for a distance of 213 m downstream from the spillway. A geophysical survey in 1996 that determined the thickness of the sediment above the undisturbed shale indicated close agreement with the earlier COE surveys near the structure, but showed the undisturbed material at higher elevations fur ther downstream.

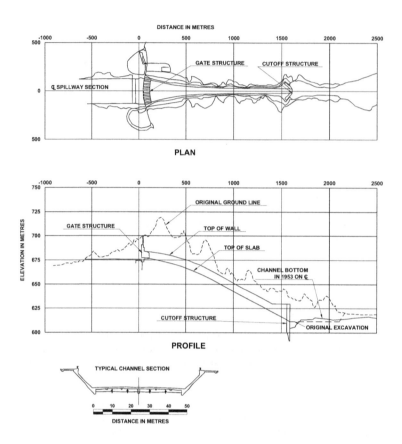

Figure 1.   Fort Peck Spillway, Missouri River, Montana.

As shown in Table 1, the maximum scour hole depth after the short duration test flows in 1946 was 4.3 m. The much longer duration flows in the following six years increased the maximum scour depth only another 3.7 m to a depth of 8.0 m. This suggests that the scouring process produced by the high velocity jet exiting the spillway approached equilibrium in a relatively short period of time.

The lateral extent of scour, however, generated by lower velocity return currents moving in the upstream direction, required considerably more time to approach equilibrium. Limited lateral scour was produced by the short test flows of 1946 with the more developed lateral scour oc-curring during the 90-day flow duration of 1947. Lateral scour progressed even further during the 90-day flow duration of 1948. During another 90-day flow duration of 1949, although at reduced flows, the lateral scour progressed moderately more. The observation that lateral scour (and also total volume) progressed over the 180-day flow duration of 1947-1948 suggests that lateral equilibrium would not be reached during the passage of a PMF runoff event.

Eroded shale deposited in a bar approximately 213 m downstream from the spillway. During a site visit to Fort Peck Dam June, 1995, a cursory determination of possible material sizes re-moved from the scour hole and deposited on the downstream bar was made by digging several holes through the bar material with a shovel to a depth of approximately 75 cm. Although the deposited material was in various stages of disintegration, discrete particles of scoured shale up to 10 cm diameter were observed, with some evidence of disintegrated larger pieces. Since sampling with this method only reached the near-surface material, considerably larger material may be present at greater depths

Table 1. Spillway Flow and Scour Hole Data.

Discharge (m³/s)	Date (year)	Duration (days)	Scour Depth (m)	Eroded Vol. (Mm³)
765	1946	< 1 Day	4.3	0.33
565 - 640	1947	90	5.6	1.15
565 - 708	1948	90	7.5	1.5
280 - 450	1949	90	6.1	1.6
425 - 565	1950-53	60 - 90	7.9	1.9
565	1975	30	Not Avail.	Not Avail.
140 - 425	1976	60	Not Avail.	Not Avail.
280 - 425	1979	30	Not Avail.	Not Avail.
113	1996	8	Not Avail.	Not Avail.

Source COE Omaha District 1994

3 MODEL STUDY

A series of model studies conducted in 1935 at a scale of 1:36 indicated a configuration with no downstream dissipator structure produced considerable downstream scour, and a comprehensive test program was then undertaken to develop a stilling basin with floor blocks to reduce the scour. However, a decision was later made to replace the stilling basin concept with a 21 m deep cutoff wall to protect against the scour. Bed material with a d_{50} of 14 mm (503 mm in prototype) was used in the tests.

A model test using a configuration relatively close to the geometry constructed in the prototype at a discharge of 7,220 m³/s was selected for use in this study. The test geometry varied from the constructed geometry as follows: (1) the channel width at the downstream end of the spillway was 39.6 m rather than the constructed 36.6 m, (2) the sidewall slope was 1H:2V rather than the constructed 1:1 slope, and (3) the downstream wingwalls simulated in the model were at 90° to the spillway centerline, rather than angled back at 135° as in the prototype. The spillway terminal slope (5.23 %) and exit elevation (613 m) were the same. The differences in exit geometry are considered negligible, particularly at the high flow of 7,220 m³/s. The constructed angled wingwall is considered to produce slightly less scour at the wall than the perpendicular model wingwalls.

The deepest scour measured after the model test was at elevation 583 m, about 76 m downstream from the spillway exit. This scour is about 30 m below the elevation of the spillway exit (613 m) and about 9 m below the base of the cutoff wall (elevation 591.6 m). The maximum scour at the cutoff wall was at elevation 599.2 m, about 7.6 m above the bottom of the wall.

4 SCOUR PREDICTION EQUATIONS

Numerous empirical scour prediction equations have been developed for both plunge pool and culvert scour. Although neither plunge pool nor culvert scour exactly simulates the scour process at Fort Peck, the availability of studies in these related applications provides a useful input to the analysis. The Fort Peck spillway flows differ from the plunge pool flows in that the Fort Peck spillway discharges below tailwater (but not deeply submerged) and at a near horizontal direction (downward slope of 5.23 %), whereas the plunge pool equations have been developed for falling jets piercing the water surface from above at steeper angles. Culvert flows, on the other hand, generally discharge into the downstream receiving channel with near horizontal directions and near the tailwater level. The shape of the culvert jet, however, being circular, differs from the rectangular jets discharging from spillways.

Mason and Arumugam (1985) have identified 31 methods for predicting scour depth for plunge pools. Seventeen of these equations relate equilibrium scour depth to discharge, head drop and particle size. The remaining 14 equations incorporate the effects of other parameters such as angle of entry, tailwater depth, air concentration, and jet dimensions. The plunge pool type equations have been summarized and discussed in detail in a literature survey on erosion (US Bureau of Reclamation, 1993). Eight of the equations were used in the initial analysis to compare predicted scour with both the model study results at high flow and the observed prototype scour at lower flows. The plunge pool equations generally have the form:

$$D = \frac{K\left(q^x \cdot H^y\right)}{d^z} \qquad (1)$$

where: D = depth of scour measured from the tailwater level; q = discharge per unit width; H = head; and d = diameter of the bed material.

The coefficient x for the flow rate is relatively constant for the various equations, ranging from about 0.5 to 0.6. The coefficient y for head varies from about 0.2 to 0.5. The coefficient z, reflecting the significance of the bed material diameter on scour, varies from 0 to 0.5.

An equation developed by Mason and Arumugam (1985) from an extensive analysis of both model and prototype observations has the form:

$$D = \frac{K \cdot q^x \cdot H^y \cdot h^w}{g^v \cdot d^z}, \qquad (2)$$

where $K = 6.42 - 3.10\,H^{0.10}$; $v = 0.30$; h = downstream flow depth (m); w = 0.15; $x = 0.60 - H/300$; $y = 0.05 + H/200$; and z = 0.10 (with an assumed constant value of d = 0.25 m recommended for all prototype materials, including shale).

Numerous relationships for predicting culvert scour have been developed. The ones considered for the purposes of this study are those by Blaisdell and Anderson (1985) and Breusers and Raudkivi (1991).

Blaisdell and Anderson's study correlates the flows from culverts that discharge freely through the atmosphere and enter the tailrace from above the water surface. They relate relative scour depth Z/D (maximum scour depth to culvert diameter) to a densimetric Froude number defined by flow velocity, the densities of the rock and fluid, and the rock size. The densimetric Froude number is sufficiently large for the Fort Peck flows, however, that the relative scour depth Z/D is constant at 7.5 for all flows. Blaisdell and Anderson also were successful in representing culvert scour contour lines with ellipses. They also introduced the concept of "beaching", which occurs at high flows where upstream directed counterflows scour the banks and transport material upstream along the periphery of the scour hole. Scour holes with beaches are excessively wide.

Breusers and Raudkivi, in the IAHR Design Manual "Scouring", have used the data of Bohan (1970), who conducted culvert tests at the Waterways Experiment Station, to recommend the following equation to predict scour downstream of culverts.

$$\frac{y_s}{D} = 0.65\left(\frac{U_0}{U_{*c}}\right)^{\frac{1}{3}} \qquad (3)$$

where y_s = scour depth below the culvert invert; U_0 = culvert velocity; U_{*c} = shear velocity determined from Shields graph.

The shear velocity is proportional to the square root of the bed material diameter and equals 45 cm/s for a bed particle size of 0.25 m. In this equation the scour depth is inversely proportional to the bed particle size to the power of 0.165. This equation is applicable to culverts that discharge near horizontally at or below the water surface. This equation is preferred over the equation of Blaisdell and Anderson since it includes the effect of particle size and more closely simulates the hydraulic entry conditions at Fort Peck.

5 DEVELOPMENT OF SCOUR PREDICTION MODEL

5.1 Overview of selected procedure

The adopted scour prediction procedure combined the use of one of the standard maximum scour-depth equations with the results from the model test conducted in 1935 during development of the initial design. The maximum scour depth equation was selected largely on its ability to closely predict the measured model scour, adjusted for material size used in the model, for a high simulated prototype flow of 7,220 m³/s and the prototype observed scour for a lesser flow of 708 m³/s. Once the standard maximum scour equation was selected, it was calibrated to the maximum scour observed in the model, but adjusted for particle size used in the model test. The calibrated equation was then used to compute the maximum scour depth for five additional flows.

In addition to the prediction of maximum scour, the location of the maximum scour and the general extent of the overall scour hole were to be estimated. The 1935 model test provided the basic dimensions for the scour hole produced by the 7,220 m³/s calibration flow. The scour hole contours were approximated with ellipses, in accordance with a procedure suggested by Blaisdell and Anderson (1991) for culvert scour. This basic scour shape, measured in the model for 7,220 m³/s, was also used for other flow rates by scaling the linear dimensions (length, width, and location of maximum scour) by the ratio of the maximum scour depth computed by the calibrated equation for maximum scour to the maximum depth for the calibration flow.

Sections derived from the computed contours were then compared with both the model results (7,220 m³/s) and prototype measurements (708 m³/s) to assess the accuracy of the prediction procedure. An additional scaling factor was introduced to improve conformity with the prototype at low flows.

5.2 Determination of calibration scour depth from model test

The maximum scour depth below the spillway lip elevation (y_s) measured in the model test was 31 m for a model particle diameter of 14 mm, simulating a prototype diameter of 500 mm, which exceeds the estimated effective diameter of 250 mm. The maximum scour is inversely proportional to the bed particle diameter to a power that can vary from zero to 0.5, depending on the scour prediction equation referenced. An arbitrary value of 1/3 was selected for the exponent in estimating the model depth of scour if a model material selected to represent a prototype diameter of 250 mm had been used. At 7,220 m³/s, the estimated depth for this smaller size of material is estimated at 39 m. This is the target calibration depth for the flow used in the model.

5.3 Selection of scour prediction equation

Eight plunge pool equations and three culvert equations were initially considered. The depth and length of the scour hole are defined in Figure 2. The plunge pool equations predict the depth from the tailwater elevation to the bottom of the scour hole (D), whereas the culvert equations predict the scour depth from the invert of the culvert exit to the bottom of the scour hole (y_s).

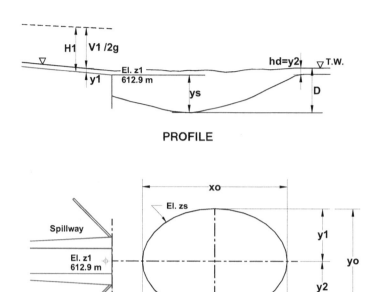

PROFILE

PLAN

Figure 2.   Scour Shape Definition.

When the equations were calibrated, the closest prediction for the maximum prototype scour depth of 8.0 m produced by the flow of 708 m^3/s was by Mason (8.2 m). The Mason equation was adopted as the basis for all subsequent analysis.

5.4   Scour hole shape

The model test scour hole contours were first approximated with two ellipses, following the procedure suggested by Blaisdell and Anderson (1991). One ellipse simulated the major downstream scour, with ellipse centers varying along the downstream projected centerline of the spillway. The second set approximated the scour adjacent to the cutoff wall produced by the upstream-directed return eddies. A judgment was made in choosing the second set of ellipses to allow for the difference in cutoff wall positions between the model and the prototype. The location of the maximum scour measured in the physical model closely approximates the location predicted for horizontally discharging culverts by Mendoza et al. (1983), who showed that the maximum depth occurred at about 0.33 of the scour hole length over a range of different conditions. The method adopted to predict scour maintains the same ratio, which agrees well with Mendoza's ratio, over the full range of discharges.

To determine the elevation and position of the scour contour lines, the vertical difference between the elevation of the spillway lip and the scour contour line (elevation z_s) was scaled proportionately from the baseline (model flow 7,220 m^3/s adjusted for particle size) maximum scour depth (z_{so}) with the maximum scour depth y_s computed from Mason's equation.

38

Scaling the horizontal dimensions (x,y) of the contour lines from the model tests by the scour depth y_s for the observed prototype scour at 708 m³/s produced scour holes that were narrower and shorter than those observed in the prototype. An additional scaling factor was introduced to ensure that the scour conditions more closely predicted the observed prototype scour. The factor was set at 1.6 for 708 m³/s, and varied linearly with the computed maximum scour depth y_s so the scale factor reduced to 1.0 at the model calibration flow rate.

5.5 Predicted scour

Representative scour predictions are plotted in Figure 3 for flow rates of 708, 3540, and 7500 m³/s. Centerline longitudinal profiles are plotted in Figure 4. The depth, length, and width of the scour holes vary from (8.9, 171, and 122 m) for 708 m³/s to (39.4, 457, and 335 m) for the maximum flow of 7,500 m³/s.

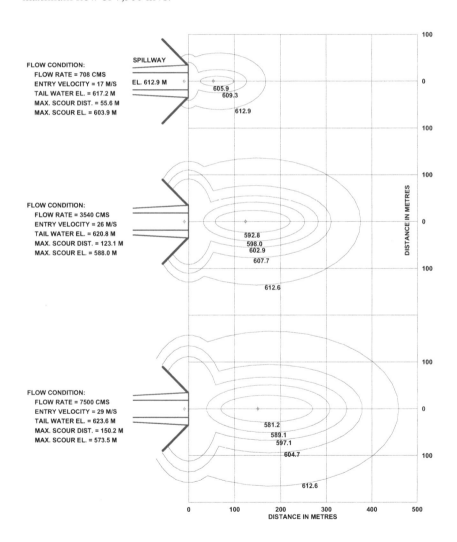

Figure 3.   Representative Scour Predictions.

39

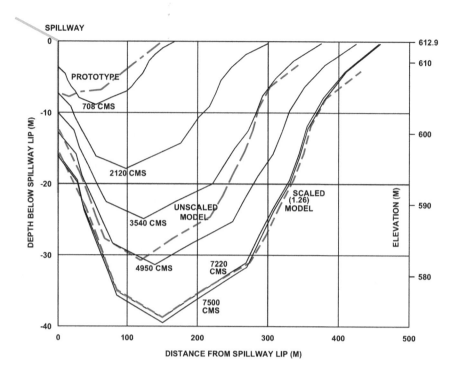

Figure 4. Scour Profiles.

5.6 Sensitivity tests

For the value of the exponent z of 0.33 assumed for the earlier analysis, the maximum scour depth for 7,220 m^3/s predicted from the model results would be reduced from 39 m for a prototype d_{50} of 0.25 m to 31 m for a d_{50} of 0.5 m and increased to 53 m for a d_{50} of 0.1 m. On the other hand, for a constant d_{50} of 0.25 m, changing z from 0.33 to 0.2 reduces the depth predicted for 7,220 m^3/s from 39 to 35 m. The predicted depths are more sensitive to estimates of prototype d_{50} than to z. The sensitivity analysis also showed that the maximum scour depths are not as sensitive to tailwater depth as to d_{50}.

5.7 Effect of time

The prototype scour records from the years 1946 to 1948 indicate rapid erosion of the bed in the area immediately downstream from the spillway. Most of the vertical erosion downstream from the spillway was observed in the survey after the 90-day flow duration of 1947, with much of the erosion occurring during the one-day test of 1946. Little additional vertical erosion occurred during the 90-day flow duration of 1948. Lateral scour proceeded much more slowly, however, with considerable additional scour occurring during 1948, even though the flow rates were approximately the same as in 1947.

Since the model material was cohesionless, the lateral extent of scour may have progressed more rapidly than if a cohesive material was used. The cohesiveness of the prototype shale is not considered a factor in resisting erosion from the high velocity jet downstream from the spillway, but is significant in resisting lateral scour.

The extent of lateral scour from the prediction model may be further than would occur during the passage of one flood event. Because of the time factor, the scour prediction may be conservative from the standpoint of lateral scour, but not for scour immediately downstream from the spillway.

6 SUMMARY

Background information used in the study included prototype maximum scour depths (8 m) measured after long duration flows of 708 m^3/s after project completion in the 1940's and model tests conducted with a 1:36 scale model in the 1930's with a configuration similar, but not an exact replica of the geometry constructed. The model predicted a prototype scour of 31 m at 7,220 m^3/s when simulating a prototype d_{50} material size of 500 mm.

Mason's maximum-scour depth-prediction equation was selected after an evaluation of many available methods. The selected model was calibrated to predict the measured model scour (adjusted for particle size) at 7,220 m^3/s, and then used to predict maximum scour depths at other flows. The calibrated Mason equation closely predicted the maximum prototype scour depth of 8 m.

The shape of the scour hole used ellipses to approximate the scour in the model, and scaling factors chosen to closely approximate prototype scour at lower flows.

Sensitivity tests indicated that the scour depth is most sensitive to estimates of prototype material bed size. Tailwater depth and the exponent z (attached to the bed material size) were less significant.

Vertical erosion downstream from the spillway occurred rapidly, whereas lateral scour of the cohesive material produced by upstream-directed return currents occurred relatively slowly.

7 ACKNOWLEDGEMENTS

The scour prediction study component, supported by the U. S. Army Corps of Engineers, Omaha District, was conducted by Northwest Hydraulic Consultants under the overall direction of R. W. Beck Inc.

8 REFERENCES

Blaisdell, F.W., C.L. Anderson. 1991. A Comprehensive Generalized Study of Scour at Cantilevered Pipe Outlets, *Journal of Hydraulic Engineering, vol. 117, No. 3, ASCE.*

Bohan, J.P. 1970. Erosion and Rip Rap Requirements at Culvert and Storm-Drain Outlets, *U.S. Army Eng. Waterways Exp. St. Vicksburg Res. Rep. H-70-2.*

Breusers, H. N. C., A.J. Raudkivi. 1991. Scouring, Hydraulic Structural Design Manual, *International Association for Hydraulic Research.*

Mason, P.J., K. Arumugam. 1985. Free Jet Scour Below Dams and Flip Buckets, *Journal of Hydraulic Engineering, vol. 111, No. 2, ASCE.*

Mendoza, C., Abt, S.R., Ruff, J.F. 1983. Headwall Influences on Scour at Culvert Outlets, *Journal of Hydraulic Engineering, Vol. 109, No. 7, ASCE.*

Wittler, R., Mefford, B., Annandale, S., Abt, S., Ruff, J. 1993. Dam Foundation Erosion, *Pre-Test Report, Release Draft.*

Slab stability in hydraulic jump stilling basins as derived from physical modeling

E. Caroni & V. Fiorotto
University of Trieste, Civil Engineering Dept, Trieste, Italy

M. Belicchi
University of Parma, Civil Engineering Dept, Parma, Italy

ABSTRACT: The paper deals with the characterization and the control of transient uplift generated by turbulent pressure fluctuations in spillway stilling basins at the base of a large gravity dam. An experimental facility, meant to measure forces and pressures acting on the slabs at the bottom in the hydraulic jump region is presented. Analyses of the sampled data with the aim to evaluate the temporal evolution of the highest uplifts forces acting on the lining are carried out. Moving from these analyses the design of the floor concrete slabs is performed.

1 INTRODUCTION

Some cases of damage experienced on spillways chute and basins, operating under flood conditions, showed the importance to define the hydrodynamic forces involved in the design of the lining of spillway chutes and stilling basins.

Most notable were the cases of the Malpaso, Tarbela and Karnafuli dams. In the former case, the 100 m head stilling basins underwent complete dislodgment of several 720 – ton slabs, each anchored with 12 steel bars. The slabs were cast in situ with expansion joints carefully peaked with bituminous caulking. The spillway discharge causing the damage peaked at one-third of the design flood tested in a 1:100 model (Sànchez Bribiesca & Capella Viscaìno 1973). In the latter case (e.g. Bowers & Tsai 1969), a discharge of about 20% the design discharge of 18000 m^3s^{-1} produced extensive damage to the chute floor over an area about 180 m wide and 23 m long. Questions have therefore been raised on whether conventional design criteria (e.g. *Design of small dams* 1977, pp. 442-446), based on steady seepage uplift, are adequate for high-head dissipation structures.

In both the Malpaso dam and the Karnafuli dam cases, it was concluded that conventional design had been correctly applied. Damage was not caused by steady uplift pressure under the chute slab nor to misalignments of the lining slabs: in both cases the severe pressure fluctuations, associated with the dissipative phenomena in hydraulic jump, were recognized as cause of the damages in the linings.

According to these evidences a design criterion for engineering practice based on the extent of the uplift induced by the severe fluctuations associated with energy dissipation in the hydraulic jump region was developed by Fiorotto & Rinaldo (1992a) on the basis of the following considerations:

– the pulsating pressures may damage the joint seal of the slabs and, through these unsealed joints, extreme pressure values may propagate from the upper to the lower surface of the slabs;
– the instantaneous difference between the total pressure acting on the upper and lower surface of the slab can reach high values. The total force related to these pressures can exceed the weight of the slab and the anchor resistance. In this context a key role is therefore played by

the instantaneous spatial structure of pressure fluctuations at the bottom of the hydraulic jump.

This criterion defines the equivalent thickness s of the slab (including the anchors contribution if present), by the relation

$$s = \Omega\left(\frac{l_x}{y}, \frac{l_x}{I_x}, \frac{l_y}{I_y}\right) \frac{p'^+_{max} + p'^-_{max}}{\gamma_c - \gamma} = \frac{F'_{max}}{(\gamma_c - \gamma)l_x l_y} \tag{1}$$

where Ω is a dimensionless coefficient related to the instantaneous spatial distribution of the pulsating pressure, γ and γ_c are the specific weight of water and concrete, respectively, and F'_{max} represents the maximum uplift force. Since the pressure time series $p(t)$ is a random stationary process (e.g. Vasiliev & Bukreyev 1967), it is convenient to use the pressure fluctuation $p'(t) = p(t) - \bar{p}$ around the mean pressure value $\bar{p}$. The p'^+_{max} and p'^-_{max} are the maximum measured pressure fluctuation values, positive and negative, respectively. The Ω coefficient depends on the ratio of the slab dimensions, l_x and l_y, to the depth of the incident flow y, and on the integral scales of the correlation function in the longitudinal and transversal directions, I_x and I_y defined as $I_j = \int_0^\infty \rho_j d_j$ where ρ_j is the pressure correlation function along the j axis direction or time. These integral scales are referred to as macroscales, to be distinguished from the microscale, which is defined as $\lambda_j \cong \sqrt{2/|\partial^2 \rho_j / \partial j^2|_0} << I_j$. The first parameter defines the time interval or space distance at which, on average, two instantaneous values of the fluctuating pressure become uncorrelated, while the second one defines the distance or time of persistency of the pressure field.

Due to non-homogeneity in the turbulence field along the flow direction, these integral scales are not constant in l_x; this variability is taken into account by the l_x/y parameter.

Determination of the reduction coefficient Ω can be performed either by direct measurement in hydraulic models of the maximum uplift forces and pressures, as

$$\Omega = \frac{F'_{max}}{\left(p'^+_{max} + p'^-_{max}\right)l_x l_y} \tag{2}$$

or, by use of the second order moments of force and pressure fluctuations, when the additional hypothesis of normality in the probability distributions can be made, as (Bellin & Fiorotto 1995)

$$\Omega = \frac{\sigma_F}{2\sigma_p l_x l_y} \tag{3}$$

where σ_F and σ_p are force and pressure standard deviations, respectively.

It must be noted that, while in the former definition, long acquisition times are needed in order to obtain the convergence, in the second case, convergence is faster.

The first approach can lead to an overestimation of the Ω coefficient because the maximum forces are correctly measured, while pressures are here measured at a single point, so that the achievement of the maximum pressure acting on the slab is not assured. The second approach is lesser dependent on the single point pressure gage position, because it is based on the second moment estimation. By converse, it can suffer from departures of the actual forces and pressure distributions from the gaussian type.

For design purposes, the choice of slab shape and dimensions is crucial, since it affects the value of the Ω coefficient, which, via theoretical analysis (Fiorotto & Rinaldo 1992a), can range from negligible values to 0.5 - 0.75. In fact:

44

a) with decreasing slab dimensions, i.e. as long as l_x/I_x and l_y/I_y tend to zero, the correlation between forces acting under and over the slab does increase, thus reducing the resulting uplift force unbalance; in the limit, with a perfect correlation of forces, the unbalance is null;

b) with increasing slab dimensions, i.e. as long as l_x/I_x and l_y/I_y tend to infinity, pressures, as measured over the slab and along its boundary, at distances larger than the directional integral scales, become progressively less correlated; since forces represent the integral of local pressure effects, the statistical independence of pressures together with the increment in the integration domain lead, in the limit, to a null unbalance;

c) with intermediate slab dimensions, for continuity reasons, a maximum unbalance is expected; from one-dimensional analysis this maximum is reached in the interval $2 < l/I < 4$, as applied to the x or y direction, according to the slab shape.

From the above considerations we can argue that the most suitable form of the slabs (for a given area) is rectangular, with width kept to the technical minimum, because Ω is much more sensitive to modifications of l_y, being $I_y \gg I_x$.

The maximum pressure fluctuations under the jump are usually related to the kinetic energy of the incident jet $v^2/2g$ by means of the pressure coefficients C_p^+ and C_p^- (Toso & Bowers, 1988, Fiorotto & Rinaldo 1992b) as follows

$$\frac{p'^+_{\max}}{\gamma} = C_p^+ \frac{v^2}{2g} \quad ; \quad \frac{p'^-_{\max}}{\gamma} = C_p^- \frac{v^2}{2g} \tag{4}$$

Experimental values of the pressure coefficient can be found in literature, and are reported depending on the spillway and stilling basin geometry, boundary layer development, Froude number and sampling duration. In particular, these coefficients increase with measurement duration.

From a physical point of view, a maximum value must exist, being the inflow energy a finite quantity. Yet, since this extreme value is unknown, and, possibly, much larger than the actual ones affecting the stilling basin, a statistical approach can be sought, aimed at defining an experimental maximum value over a sampling period. In this context, a sampling duration, larger than the expected duration of floods evacuated by the spillway, can be seen as a safety factor.

While the sampling duration can be seen as a safety factor on the estimation of maximum pressure fluctuations, a further safety coefficient must be applied to the Ω coefficient, coping for model approximation and scale effects.

In a physical model analysis, the uplift force $F'_{\max}$ is directly evaluated, thus avoiding the estimation of $p'_{\max}$ and Ω. Yet, a safety coefficient must be imposed as well, with reference to the sampling duration and model approximation and scale effects.

The direct approach does not take into account the cavitation phenomena that can occur in the hydraulic jump for high velocity of the incident flow at the base of large dam spillway (Hager 1992). This phenomena define a lower limit to the negative fluctuating pressure; assuming $\bar{p}/\gamma \approx$ the sequential depth in the hydraulic jump, the limit lower pressure becomes equal to $p'^-_{\max} = p_v$ where p_v represents the fluid vapor pressure,.

Cavitation is not reproduced in the hydraulic models where, according to the Froude similarity, pressures are scaled with the model geometric scale ξ while the fluid vapor pressure, that depends of the fluid characteristics, is practically the same as in the prototype. Thus, reaching the condition $p'^-_{\max} \geq p_v \xi$ in the model, means that in the prototype cavitation effects might occur. Neglecting this effect leads to an excessive value of the slab equivalent thickness because $p'^-_{\max}$ in equation 1 is overestimated. In this case equation 1 becomes

$$s = \Omega\left(\frac{l_x}{y}, \frac{l_x}{I_x}, \frac{l_y}{I_y}\right) \frac{p'^+_{max} + p_v}{\gamma_c - \gamma} \tag{5}$$

The design criterion expressed by Equation 1 and 5 is correct when the slab stability is ensured only by its weight, but, as long as a fraction of the uplift force is balanced by anchors, the dynamic behavior of the system (slabs + anchors) must be taken into account. In this context, not only the intensity, but also the temporal evolution of the pressure field at the bottom of the hydraulic jump plays a relevant role.

In the anchored slab case, the total uplift force is balanced by the sum of the concrete weight and the steel bars resistance; application of Equation 1 can be made according to

$$s = \frac{\Omega}{(\gamma_c - \gamma)} 2p'_{max} = s_s + \sigma_a \frac{A_{st}}{(\gamma_c - \gamma)} \tag{6}$$

where s_s is the design concrete slab thickness, σ_a the admissible tension in the steel bars and A_{st} the steel area per slab surface unit.

Equation 3 is based on a static force balance, yet, to evaluate the real slab-anchor system behavior, the concrete slab inertia and the steel bar elastic properties must be taken into account within a dynamic analysis.

Theoretical analysis (Fiorotto & Salandin 2000) pointed out that: i) the equivalent thickness criterion (Eq. 3) can underestimate the steel area needed for the anchors ii) to maintain the maximum stress within the admissible values in every case, the steel area computed by the equivalent thickness criterion (Eq. 3) must be doubled. In this case, the design criterion must meet the requirements on the admissible displacement of the slab.

The application of results presented in literature, related to the C_p and Ω coefficients, can lead to overestimation in the equivalent slab thickness, as can be guessed considering the difficulty in their extrapolation to practical cases, where the overall geometry can significantly depart from the cases examined in literature.

In applications to large dams, where consistent economy in project costs can be expected from an accurate evaluation of slab dimensions, a direct experimental assessment can prove of some efficiency.

The aim of this work is to present an experimental facility, meant to measure forces and pressures acting in a stilling basin physical model in order to design the protection slabs.

The paper is organized as follows. In the first part the experimental burden is described and results are analyzed. In the second part the design of the anchored slabs at the stilling basin bottom is presented.

2 THE DAM PHYSICAL MODEL AND EXPERIMENTAL DETAILS

A large gravity dam was modeled in a 1:40 geometric scale, according to the Froude similarity.

The dam crest elevation is 342.00 m a.s.l. while the spillway crest is at 337.80 m a.s.l.. The stilling basin bottom is at 290.45 m a.s.l. so that the spillway height is 47.35 m at a slope of 1:0.8. The stilling basin is 46.75 m long, 80 m wide, sloping to the elevation of 290.00 m. With reference to the previous considerations on the Ω coefficient, and technical and economical constraints, a rectangular slab with dimension $l_x = 8$ m and $l_y = 4$ m was chosen. At the downstream end of the stilling basin, a weir, with crest at 293.75 m a.s.l., controls the hydraulic jump.

For a design discharge of 800 m^3s^{-1} the head is 2.7 m, which, for a fall of 50 m, yields a toe velocity (Chow 1959) of 24 ms^{-1}. Hence, for a specific discharge of 10.0 m^2s^{-1}, a flow depth of 0.42 m is computed, corresponding to a Froude number equal to 12. From the graphs of Bellin & Fiorotto (1995), which range up to $l_y/y_1 = 8$ and $l_x/y_1 = 10.4$, an extrapolated value for $l_y/y_1 = 9.6$ and $l_x/y_1 = 19.2$, can be obtained as $\Omega = 0.15$; according to Equation 1, assuming a value of C_p^+ and C_p^- equal to 0.7 according to data by Toso & Bowers (1988), an equivalent thickness of the slab equal to 3.5 m is obtained.

Obviously, this value is to be intended as a first rough estimation, and, at least for important structures, a more accurate evaluation can be obtained by use of physical models. With reference to the Italian regulations on large dam construction, the use of physical modeling is suggested in order to check the hydraulic behavior of the structure. In this context, an investigation on the stilling basin slab stability was conducted.

The experimental setup is shown in Figure 1. An aluminum frame 1000 mm long, 500 mm wide and 25 mm thick with a central rectangular hollow, 10 mm deep, was inserted in the stilling basin bottom. Inside the hollow, a movable aluminum slab (100×200 mm) 8 mm thick was cast with the upper face at the same level (up to O(10^{-5}) m accuracy) of the stilling basin flume (Fig. 2).

The dimensions of the hollow were such as to leave a gap of about 2 mm along the slab sides. Through this gap, the fluctuating pressure at the bottom of the hydraulic jump propagates in a 2 mm thin water layer under each slab. This propagation takes place with negligible friction effects in the reduction of the uplift pressures, according to the theoretical analysis as applied to real cases (Fiorotto & Rinaldo 1992a); anyway, by neglecting the friction damping, a safer result is obtained.

In the comparison between model and prototype conditions, the presence of different materials (aluminum in the model and concrete-ground contact in the prototype) can induce differences in pressure propagation celerity; however, this effect is negligible, because the time microscale of the pulsating pressure λ_t due to large scale eddies is larger than the propagation time T, that is $T/\lambda_t \ll 1$, both in the prototype and in the model.

In fact, the celerity of pressure propagation in the prototype ranges in O(10^2-10^3) ms^{-1}, while the persistence time of the pulsating pressure is O(1) s (Bowers & Tsai 1969; Toso & Bowers 1988); in comparison, for a slab with characteristic dimension of 10 m, the propagation time is O(10^{-1}-10^{-2}) s.

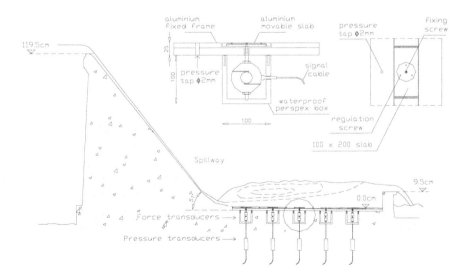

Figure 1. Experimental setup of a large dam stilling basin model.

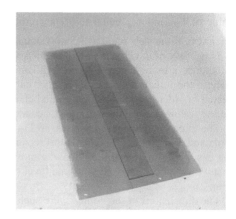

Figure 2. View of the experimental installation and particular of the test area.

The same happens in the model, where the pressure propagation celerity is $O(10^3)$ ms^{-1}, that is of the same order of magnitude of the prototype, while persistence time and slab dimensions are scaled according to the Froude similarity. By consequence, indicating with ξ ($\xi < 1$) the model geometric scale, the propagation time T scales approximately with the slab dimensions $l_{model}/l_{prototype} = \xi$ while the turbulence time macroscale scales with time $t_{model}/t_{prototype} = \sqrt{\xi}$, so that $(T/\lambda_t)_{model} < \sqrt{\xi}(T/\lambda_t)_{prototype} \ll 1$.

For this reason, the model and the prototype behave similarly, that is, the pulsating pressures are not significantly damped and the resulting force due to the uplift pressures can be computed by the average of the instantaneous pressures acting at the joints.

The alignment of the upper face of the movable slab with the basin bottom is a crucial problem because even a small inaccuracy can alter the pressure field around the slab, thus inducing errors in the measured forces. For this reason, the movable slabs were fixed by means of three micrometric regulation screws and a fixing screw, to a circular plate coupled to the force transducers (Fig. 1).

Both the movable slabs the circular coupling plate and the joint (hollowed) were built in aluminum, as light as possible, compatible with the requirements of resistance and stiffness, to minimize the inertial effect on force measurement.

Each force transducer (Fig. 1) was placed in a rigid waterproof Perspex box, with 10 mm thick walls, fixed to the aluminum frame. The joint connecting the movable slabs to the force transducers have a diameter smaller than the hole in the frame to avoid friction effects on force measurement. As a consequence the water in the Perspex box is directly connected with the water film under the movable slabs. Thus a rigid waterproof container is needed, to prevent damping phenomena in underpressure propagation and vibration in force transducers, which are fixed to the box bottom.

Force transducers TS100 of AEP Transducers with a sensitivity less than 0.1 N, and a response time lesser than the microscale time of the pulsating forces were used. The choice of this type of force transducer was conditioned by the need of coping with measurement accuracy, hardware robustness compatible with use in large hydraulic models, waterproof sealing (IP68) to operate in submerged conditions, and long duration dynamic applications. The signal was amplified and conditioned via TA4 analog transmitters (AEP Transducers), with a frequency response of 1 kHz.

Five pressure taps were inserted on the basin bottom, aligned with the center of each slab at a distance from the spillway base equal to: Tap 1, 0.1 m, Tap 2, 0.3 m, Tap 3, 0.5 m, Tap 4, 0.7 m, Tap 5, 0.9 m. The pressure taps have a diameter of 2 mm and were connected to the transducers by a rigid tube of 4 mm internal diameter. Pressure transducers of type Foxboro FPT adjusted in the range 0 - 70 kPa, with a response time (10-90%) lesser than 1 ms, that is, lower than the microscale time of the pulsating pressure (Abdul Kader & Elango 1974), were adopted.

A computer was linked to the transducers via a 64 channel analog-digital board United Electronic Instruments PD2-MF-64-400/14H.

Sampling was performed by means of the Dasylite code by Dasytech USA, implemented on a Pentium II PC; data were stored in hard disk to perform further computations.

The inflow discharge was measured by an induction flow meter MUT 2200 316 L by Automazioni Industriali, Padua, Italy, with an instrumental accuracy lesser than 0.2% and significant digits of 0.1 ls^{-1}.

3 EXPERIMENTAL RESULTS

Hydrologic flood frequency analysis indicates a 1000 year return period flood estimate varying according to the probability law and the estimation method being employed. As an average, a discharge rate of 600 m^3s^{-1} is found, while a safer estimate of 800 m^3s^{-1} is indicated as design discharge.

Consequently, two test runs were performed, whose characteristics are reported in Table 1.

For both tests, data were achieved at a 150 Hz frequency rate, according to the Nyquist criterion, for a 24 hour duration, corresponding to a prototype flood duration of about 6 days.

The toe of the hydraulic jump is located for the maximum discharge at the spillway base (Fig. 3), while, decreasing the discharge, the toe of the jump moves upstream on the chute.

Table 1. Test characteristics.

		Test A		Test B	
		model	prototype	model	prototype
discharge	(m^3s^{-1})	$79 \cdot 10^{-3}$	800	$59.3 \cdot 10^{-3}$	600
initial velocity	(ms^{-1})	3.8	24	3.5	22
initial depth	(m)	0.010	0.42	0.008	0.34
initial kinetic head	(m)	0.74	29	0.63	25
Froude number	(-)	12.	12.	12.	12.
Reynolds number	(-)	$40 \cdot 10^3$	$10 \cdot 10^6$	$30 \cdot 10^3$	$7.5 \cdot 10^6$
run duration	(hours)	24	152	24	152

Figure 3. Views of the hydraulic jump for the maximum discharge

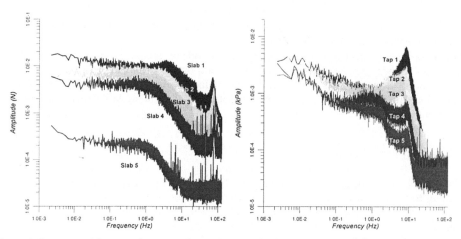

Figure 4. Spectrum of forces and pressure fluctuation amplitude, Test A, 800 m³s⁻¹

In Figure 4 forces and pressures amplitude spectra are respectively reported. From the figure one can observe that, as compared to pressures, forces have a broader spectrum. The analysis of the amplitude spectrum suggests that the fluctuation energy is mainly concentrated at the lower end of the spectrum with a small energy spreading toward frequencies larger than the characteristic pressure frequencies.

In fact, from Figure 4, the dominant frequencies of the pressure spectrum are below 30 Hz, while for the force spectrum the decay is slower, with dominant frequencies below 70 Hz, so that the choice of a sampling frequency of 150 Hz is consistent with the Nyquist criterion.

The spreading of energy could be a consequence of a weak non-linearity in the physical processes that control the propagation of underpressure below the slabs. Since most part of energy is located at small frequencies, the processes that create the uplift forces are mainly controlled by the dynamics of large eddies which are scaled according to the Froude law (Vasiliev & Bukreyev 1967).

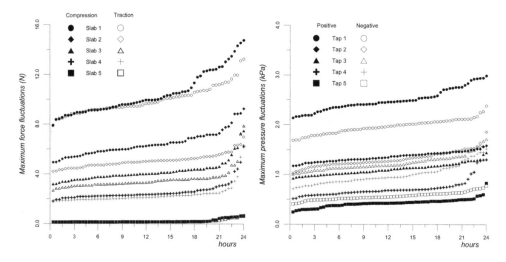

Figure 5. Effects of run time on positive and negative force and pressure peak deviations from mean, for Test A.

50

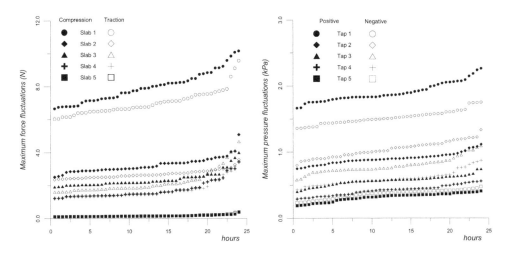

Figure 6. Effects of run time on positive and negative force and pressure peak deviations from mean, for Test B.

As related to the slab design criteria (Eqs. 1-6), the maximum pressure and force fluctuations are of interest.

In Figures 5-6, the maximum value of positive and negative force and pressure fluctuations are reported for Test A and Test B, respectively. The time series were subdivided into contiguous equal duration sub-periods (0.5 hours, of the same order of magnitude of a single flood peak duration, 3-4 hours in reality), and force and pressure extremes were detected and sorted in increasing order.

From the figures, one can note that pressure and force distributions increase with the run time, so that it is very difficult to define a possible upper limit. Moreover the positive forces (compression) are larger than the negative ones (traction) and the maximum values of force and pressure decrease, increasing the distance from the jump toe, Table 2. These maximum values measured on a 24 hours run time are equal to 8 – 10 times the standard deviation.

Table 2. Test results

Slab/Tap number	Test A					Test B				
	1	2	3	4	5	1	2	3	4	5
x/y_1	9.6	28.8	48.0	67.3	86.5	11.9	35.6	59.3	83.1	106.8
p'^{+}_{max} (kPa)	3.00	1.58	1.44	1.30	0.83	2.26	1.11	0.74	0.56	0.41
p'^{-}_{max} (kPa)	2.38	1.85	1.71	1.52	0.88	1.75	1.33	1.09	0.87	0.48
σ_p (kPa)	0.41	0.25	0.20	0.13	0.09	0.33	0.17	0.12	0.10	0.09
F^{+}_{max} (N)	14.8	9.3	7.9	6.3	0.7	10.2	5.1	4.0	3.4	0.4
F^{-}_{max} (N)	13.2	7.0	7.5	6.2	0.6	9.6	3.6	4.7	3.4	0.4
σ_F (N)	1.67	0.90	0.63	0.38	0.03	1.13	0.53	0.35	0.27	0.03
Ω (Eq. 2)	0.12	0.13	0.11	0.10	0.02	0.11	0.10	0.11	0.10	0.02
Ω (Eq. 3)	0.10	0.09	0.08	0.08	0.01	0.09	0.08	0.08	0.07	0.01

With reference to the pressure fluctuation, the C_p values are 0.41, 0.24, 0.26, 0.21, 0.11 for the 800 m^3s^{-1} discharge, passing from Tap 1 to Tap 5, and 0.36, 0.21, 0.17, 0.14, 0.08 for the 600 m^3s^{-1} discharge.

These values are of the same order of magnitude of those ones computed by Fiorotto & Rinaldo (1992b) and Toso & Bowers (1988), taking into account the different characteristics of the inflow conditions and of the hydraulic jump position. In fact, the peak values of C_p occur at the beginning of the stilling basin, since the toe of the hydraulic jump is located on the spillway chute, upstream of the stilling basin (Fig. 3).

The maximum value of the Ω coefficient ranges between 0.10-0.13. This value is close to the value 0.15, which was extrapolated as a first estimate from the experimental results by Bellin & Fiorotto (1995).

The Ω reduction coefficients, as estimated via Equation 2 by means of the maximum value of force and pressure, are 20 – 30 % larger than the ones computed via Equation 3, by means of force and pressure standard deviations. The first approach leads to an overestimation of the Ω coefficient because the maximum forces are correctly measured, while the measured maximum pressure acting on the slab is not assured.

On the contrary, the second approach (Eq. 3), based on the statistical moments estimation, could underestimate Ω if the Probability Density Function (PDF) of pressure and force depart from the gaussian type.

Previous works (e.g. Fiorotto and Rinaldo 1992b) treated the PDF of the pressure fluctuations under hydraulic jump. These works pointed out that:
a)   in the first zone of the jump, the positive fluctuating pressures are relatively more frequent than the negative fluctuating pressure of the same value;
b)   increasing the downstream distance, the positive fluctuating pressures show about the same frequencies as the negative pressure fluctuations, and the PDF is practically gaussian;
c)   at the end of the jump, where the skewness parameter is negative, the negative pressures are more frequent than the positive fluctuations of the same value.

In Figure 7, the PDF's of force fluctuations are shown, as compared to the normal distribution. From the figure one can note:
a)   in the first zone of the jump (Slab 1) the forces PDF departs from the gaussian behavior while increasing the force intensity, even if the positive fluctuating forces show about the same frequencies as the negative forces fluctuations of the same value;
b)   increasing the downstream distance, the positive fluctuating forces show about the same frequencies as the negative forces fluctuations, and the PDF tends to become practically gaussian.

A difference in PDF between force and pressure is found out in the first part of the jump where the positive pressures are relatively more frequent than the negative ones of the same value, while the forces distribution is practically symmetrical.

This fact is congruent with the process of force generation: in fact i) the negative (traction) uplift forces are generated by positive fluctuating pressures propagating through the joints from the upper to the lower surface of the slab, while a persistence of a negative fluctuating pressure occurs on the upper surface of the slab; ii) the negative (compression) forces are generated by the negative fluctuating pressure propagating through the joints from the upper to the lower surface of the slab while a persistence of a positive fluctuating pressure occurs on the upper surface of the slab.

It is manifest that force is generated by the spatial distribution - over the slab or at the joints - of the positive and negative fluctuating pressures and that these ones have practically the same role and importance. As a consequence a symmetry in force PDF must be expected.

Increasing the distance from the hydraulic toe, pressures and forces PDF's both tend to become practically gaussian, so that the estimate of the reduction coefficient Ω may be computed, moving from this assumption, via Equation 3.

52

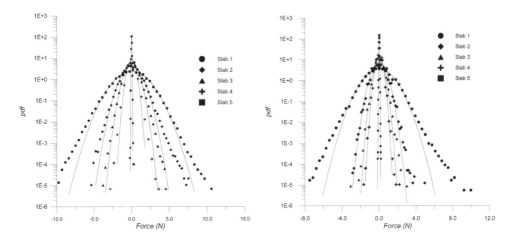

Figure 7. Force fluctuation PDF, as compared to the normal distribution: Test A (left) and Test B (right)

The analysis of Table 2 shows a small percentage variation (15%) in the reduction coefficient estimates, when moving from one slab to the other in the hydraulic jump area, where forces and pressures assume significant values for practical purposes.

This same pattern is evidenced irrespectively of the method used to derive the reduction factor, by the maximum value of forces and pressures (Eq. 2) or by forces and pressures standard deviations (Eq. 3).

This fact is a guarantee of the goodness of the Ω estimation via force and pressure standard deviations, which proves statistically more robust, with shorter sampling duration. The ratio between the two values, as computed via Equation 2 and Equation 3, equal to $1.2 - 1.3$, could be assumed as a safety factor to apply to the reduction factors when these are estimated according to Equation 3.

4 DESIGN CRITERIA

These experimental results allow to define the equivalent thickness s that must be assigned to the slab to assure the stability of the lining in the spillway stilling basin.

With reference to the maximum discharge of $800 \text{ m}^3\text{s}^{-1}$ the minimum fluctuating pressure in the prototype is equal to $2.38 \cdot 40 = 95.2$ (kPa) lesser than the water vapor pressure ≈ 100 (kPa), so that cavitation effects are not taken into account.

In these cases, the equivalent thickness can be computed via Equation 1 directly, using the fluctuating forces as measured in the model. Computation of the equivalent thickness is reported in Table 3, assuming a specific weight for concrete and water equal to 25 and 10 (kN m^{-3}) respectively.

Table 3.- Computation of the equivalent thickness

Slab		Test A					Test B				
		1	2	3	4	5	1	2	3	4	5
F'_{max}	(kN)	845	448	480	397	38	614	230	301	218	26
s	(m)	1.8	1.0	1.0	0.9		1.3	0.5	0.6	0.5	

Assuming a concrete thickness of the slabs equal to 1.2 m, this guarantees, with reference to the design discharge, the slab stability with a safety factors greater than 1.2 for Slabs 2 to 4.

For Slab 1, there are two alternative possibilities: to increase the concrete thickness up to 2.2 m or, adopting the same thickness of the other slabs, to anchor the slab to the bedrock.

In this latter hypothesis, assuming an admissible tension in steel bars equal to 24 kN cm^{-2}, Equation 6 gives a value of steel area per slab surface unit $A_{st} = 0.65$ cm^2 m^{-2}.

This value must be doubled in order to take into account the dynamic effects; thus a design steel area per slab surface unit equal to 1.3 cm^2 m^{-2} is obtained.

For a slab area equal to 32 m^2, the total steel area required is equal to 42 cm^2 per slab, as provided, for instance, by 6 Ø 30 steel bars.

The safety coefficient here employed, 1.2, could be considered inadequate, as compared to values suggested in literature, but it was evaluated with the maximum design discharge (Test A); for the more likely discharge with 1000 year return period (Test B), the safety coefficient increases up to more than 1.7.

REFERENCES

Abdul Kader M.H. & Elango K. 1974. Turbulent pressure beneath a hydraulic jump. *Journal of Hydraulic Research* 12(4).

Bellin A. & Fiorotto V. 1995. Direct dynamic force measurement on slabs in spillway stilling basins. *Journal of Hydraulics Engineering, ASCE* 121(10).

Bendat J.S. & Piersol A.G. 1971. *Random data: analysis and measurement procedures*, New York: J. Wiley & Sons.

Bowers C.E. & Tsai F.H. 1969. Fluctuating pressure in spillway stilling basins. *Journal of Hydraulics Division, ASCE* 95(HY6).

Chow V.T. 1959. *Open-channel hydraulics*. New York: Mc Graw Hill.

Design of small dams. 1977. Washington, D.C.: Water Resources Tech. Publications.

Fiorotto V. & Rinaldo A. 1992a. Fluctuating uplift and linings design in spillway stilling basins, *Journal of Hydraulic Engineering, ASCE* 118(4).

Fiorotto V. & Rinaldo A. 1992b. Turbulent pressure fluctuations under hydraulic jumps, *Journal of Hydraulic Research* 30(4).

Fiorotto V. & Salandin P. 2000. Design of anchored slabs in spillway stilling basins, *Journal of Hydraulic Engineering, ASCE* 127(7).

Hager W.H. 1992. *Energy dissipators and hydraulic jump*. Dordrecht: Kluwer.

Sànchez Bribiesca J.L. & Capella Viscaìno A. 1973. Turbulence effects on the lining of stilling basins, *Proceedings of XI Int. Congress on Large Dams Vol.2*. Madrid.

Toso J. & Bowers E.C. 1988. Extreme pressure in hydraulic jump stilling basins, *Journal of Hydraulics Division, ASCE* 114(8).

Vasiliev O.F. & Bukreyev V.I. 1967. Statistical characteristic of pressure fluctuations in the region of hydraulic jump. *Proceedings of XII IAHR Congress Vol.2*. Fort Collins, Co. (USA).

Local rock scour downstream large dams

R.A. Lopardo
Instituto Nacional del Agua (INA), Ezeiza, Argentina

M.C. Lopardo
Instituto Nacional del Agua (INA), Ezeiza, Argentina

J.M. Casado
Instituto Nacional del Agua (INA), Ezeiza, Argentina

ABSTRACT: This paper draws some examples from research works carried out in Argentina on local scour downstream from ski jump structures and from hydraulic jump energy dissipators.
An equation of local scour downstream from flip buckets, intentionally oversimplified so as to be of use as an initial scour estimate, is presented. Knowledge of unit discharge and fall height from reservoir level to tailwater level is required. It should be pointed out that during the design stage many parameters are unknown, such as the size of blocks formed by rock fracture at different depths near the jet impact. For preliminary calculations, this equation shows acceptable performance of data furnished by other authors, including data that was published after the equation was formulated.
The river bed is formed by local erosion downstream from hydraulic jump energy dissipators in large flatland dams, where rocks undergo severe pressure fluctuations. This problem was analyzed using fluctuating pressures (statistic amplitude and frequency values) in the base of a hydraulic jump stilling basin, downstream from a continuous end sill, taking into account amplification of the process and the trend to induce important depressions. When the riverbed is made up of granular material, the end sill collaborates for a controlled local scour near the structure. If the riverbed is composed of large rocks, prototype experimental data demonstrates that an end sill is not a good solution. It increases flow fluctuations and concentrates the turbulent energy around a dominant frequency, removing bigger blocks and favoring local scour.
A methodology to calculate the weight of eventual protecting blocks or anchorage bars to avoid removal is proposed. It takes into account the incident Froude Number of the jump and the relationship between the block's width and length. The authors propose an unconventional approach to estimate maximum local scour depth downstream of stilling basins with baffled piers (for forced energy dissipation) and not sufficient tailwater depth.
The critical flow condition over the end sill of an energy dissipater generates a new hydrodynamic condition. The accelerated flow acts on the river bottom as a submerged jet, as observed for the hydrodynamic configuration of ski-jump spillways, but with very low hydraulic head.

1 INTRODUCTION

This paper refers to experimental research on local scour due to high velocity jets and takes into account hydraulic differences in high chute spillways and hydraulic problems associated with low ogee spillways.

First, the authors use an oversimplified equation for local scour downstream from ski-jump spillways, applying new prototype data from other authors. Second, a proposal of scour depth estimation on rock riverbed downstream from stilling basins due to macroturbulent fluctuations is presented. Finally, the authors present an unconventional methodology to calculate local scour downstream from stilling basins without normal tailwater depth based on high velocity flow data.

2. EQUATION FOR LOCAL SCOUR DOWNSTREAM FLIP BUCKETS

An intentionally oversimplified formula to estimate the maximum depth of scour downstream from ski-jump spillways was presented by Chividini et al. (1983).
This equation can be expressed as:

$$y/\Delta H = K\, Z^{*0.5},$$

where y is the maximum depth of scour (Figure 4), ΔH is the fall distance between the reservoir level to the tailwater level, and Z^* is called the "fall number":

$$Z^* = q/(g\, \Delta H^3)^{0.5},$$

where q is the unit discharge and g is the acceleration due to gravity.

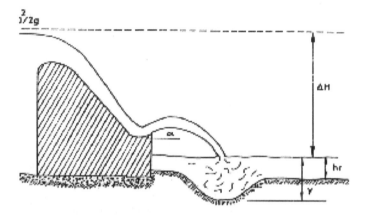

Figure 1. Ski-jump erosion notation

The equation was developed using laboratory tests performed by the authors, sixty-six experimental laboratory tests performed by other authors (with maximum scour depth standard deviation of 18%) and seventeen prototype results (with a 26% standard deviation). The validity field for this expression for the scour of no cohesive bed materials and energy losses coefficient in the spillway chute (ratio between the theoretical and actual velocity in the ski-jump section) over 0.75.
The equation developed by INCYTH only requires knowledge of the unit discharge and the fall height from the reservoir level to tailwater level. It should be noted that during the design stage many parameters, such as the size of blocks formed by fracture of the rock at different depths near the jet impact, are unknown. Performance of this equation is acceptable for prototype preliminary calculations, as demonstrated by Mason and Arumugam (1985).
The advantage of INCYTH's equation is that a simple preliminary calculation with an excellent mean performance can be made, even if the rock size is unknown (it is very difficult to estimate this parameter before the dam is built).
Lopardo and Sly (1992) demonstrated the validation of this equation by means of its comparison with depths of scour measured in prototype for other authors after the publication of the INCYTH equation.
These results were published by Balloffet (1987) for the Tarbela Dam in Pakistan, by Riedel (1989) for the Colbún Dam in Chile, by Oliveira Lemos and Matias Ramos (1984) for the

Cabora Bassa Dam in Africa, and by Keming et. al. (1987) for more than twenty scour prototype data for Chinese dams.

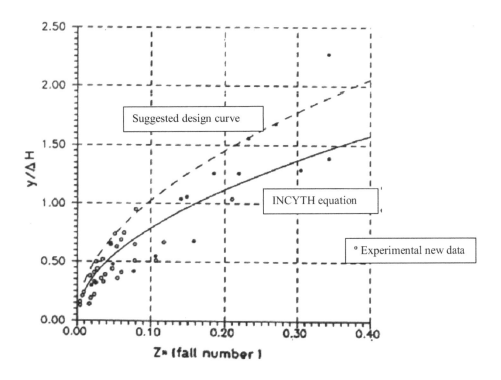

Figure 2. Comparison of INCYTH equation and new prototype data (Lopardo & Sly, 1992)

As shown in Figure 2, INCYTH's equation is always a good mean solution for these new prototype data. Nevertheless, due to safety conditions required by engineering design, it is deemed convenient to include a safety coefficient to cover 99% of existing prototype data. This coefficient should be approximately 1.3. INCYTH's equation with this safety coefficient is also presented in Figure 2 as a "suggested design curve".

3. LOCAL SCOUR ON ROCKY BEDS DOWNSTREAM FROM STILLING BASINS

Energy dissipators of large flatland dams usually consist of hydraulic jump stilling basins. The internal flow of a hydraulic jump is essentially macroturbulent, showing severe random pressure and speed fluctuations. Incident kinetic energy has a transformation along the jump and is converted to potential energy. It also generates a fluctuating energy, which is transported by different scale vortices. This energy is gradually dissipated downstream from the jump.

Rapidly variable pressure fields in space and time tend to enlarge existing bed discontinuities or create new ones when composed of rock. Fluctuating pressure propagation inside the rocky structure breaks it into minor pieces. Alternative forces induced by pressure fluctuations pull some blocks out of the bed. Drainage among the blocks reduces the fluctuation amplitude. The cavity is enlarged by ascending streams generated in the area where the jet hits against the bed.

Water pressure is highly variable along the rock-water interface, being higher in certain zones and lower in others. When the size of rocky blocks increases, the space-time correlation of fluctuating pressures tends to be reduced. A strong ascending instantaneous force will be generated if there is simultaneity of actions. Blocks should have reduced dimensions, turbulence should be of large scale or both causes should present together There is clear evidence that macroturbulence largely exceeds conventional stilling basins length. Turbulence intensity decay began at twice the basin length (basins designed using classical jump length).

The macroturbulent flow inside a hydraulic jump stilling basin is responsible for the existence of strong pressure fluctuations. Many papers have been written warning about its highly destructive nature (Lopardo et al, 1987). Structural discontinuities help to amplify severe pressure fluctuations. Also, spectrums have a tendency to concentrate energy around a dominant frequency. Forced hydraulic jumps can be produced by inserting a small wall element in the laboratory canal. The wall consist of a despicable thickness situated at a distance x_0 from the jump start, of height h_b and negligible influence of viscous forces.

The expression to calculate a non-dimensional coefficient with the root mean square fluctuating pressure amplitudes ($\sqrt{p'^2}$) is:

$$C'_p = \sqrt{\frac{p'^2}{\rho V_1^2 / 2}}$$

$$C'_p = C'_p(x/h_1, x_0/h_1, h_b/h_1, F_1)$$

where Section 1 is the upstream section of the jump (where it begins) and ρ = fluid specific mass, h_1 = vertical depth at Section 1, V_1 = velocity at Section 1, and F_1 = Froude Number at Section 1. This expression has been obtained by a dimensional analysis.

There are several specific cases of stilling basins that are shorter than the jump's length with end sills. Fluctuating pressure results have been obtained for most critical situations. Experiments were performed on a wide set of flows between $3 < F_1 < 7$. Pressure transducers where located on the end sill and downstream on the rocky-simulated bed. The physical model simulated Yacyreta's main spillway stilling basin. The end sill at this dam, which is located on the Paraná River (Argentina - Paraguay), is neither vertical nor high.

It can be stated that the fluctuating pressure coefficient has a strong decreasing trend in connection with Froude number at Section 1. It goes from $C'_p = 0.09$ for $F_1 = 3$ to $C'_p = 0.015$ for $F_1 = 7$. Strouhal Number (S_p) versus Froude Number at Section 1 has also a similar decreasing tendency for all transducers.

Another research work developed for the Yacyreta stilling basin (Lopardo & Lapetina, 1997) registered a large quantity of fluctuating pressure amplitudes and frequency data on the basin's floor. Among the many values reported for the intakes located downstream from the end sill, it is worth mentioning the following: C'_p = fluctuating pressures coefficient, $p'_{0.1\%}$ = negative values of instantaneous pressures with 0.1% probability of being trespassed, f_p = dominating frequency and f_z = zero crossing frequency.

Stilling basins with an end sill that were designed on rocky river beds using conventional criteria are 60% of the hydraulic jump's length. It should be noted that this configuration can be interpreted through this elementary analysis using a two-dimensional barrier. It can thus be concluded that end sills are not useful structures if the river bed downstream from the basin is not formed by granular materials. Blocks that are removed by the flow give rise to an irreversible situation that favours further erosion.

Experimental work was carried out to determine maximum uplift forces on bed blocks of large dimensions downstream from a two-dimensional barrier. Instantaneous values of actions are obtained by performing a discrete integration of "n" pressure values measured simultaneously using pressures gauges and a computational storage system. Force is define as:

$$F'_{ij} = \Sigma_i \, p'_{ij} \, d\Omega_i \,,$$

where p'_{ij} = fluctuating pressure registered by sensor "i" at time step "j"; i = sensor number from 1 to n; n = total number of sensors; j = data number registered from 1 to N; N = total number of measurements (time steps); and $d\Omega_i$ = area assigned to sensor "i".

Parameter j represents time but in a discrete manner because sampling frequency is constant throughout acquisition time. A maximum of 16384 force data are obtained throughout the sampling period. Statistical indicators for mean and extreme fluctuating force values can be obtained from the recorded data file. The following non-dimensional coefficient (similar to the concept of C'_p) can be defined as:

$$C'_F = \frac{F'_{(rms)}}{1/2\, \rho U_1^2\, ab}$$

where "a" and "b" are the dimensions of the calculation slab. The experiments were conducted at Section 1 using depths higher than 3 cm and Reynolds Number over 100,000. In this paper only the results obtained for $F_1 = 4$ and $x/h_1 = 26$ will be used for further calculations.

$a/h_1 = 3$ $C'_F = 0.021$
$a/h_1 = 4$ $C'_F = 0.026$
$a/h_1 = 5$ $C'_F = 0.017$

Estimates were made of the dimensions of basaltic rock blocks that could be removed by the flow downstream from the Salto Grande Dam's stilling basin (the dam is on the Uruguay River between Argentina and Uruguay). An exigent condition that really occurred was taken as the basis for comparison: $q = 70.87$ m^2/s, $F_1 = 4.17$, l = 64 m (basin length), $h_1 = 3.09$ m, $L_r = 82$ m (jump length), $U_1 = 22.96$ m/s, $x/h_1 = 26.5$ (non-dimensional position).

Results obtained are enough to calculate coefficient C'_F through root mean square force $F_{(rms)}$. C'_F is not a good indicator of extreme instantaneous force values. Using data registered for Froude number equal to 4 with fix barrier experiments, a relationship between uplift extreme 0.1% and mean values has been found. Extreme values are 3.07 to 3.20 times higher than root mean square values. The objective of these calculations is to obtain an estimate of minimum rocky block size to assure stability. For these working conditions flow can lift a block up to 1.70 m thick, 4 m length and 3 m wide. This result agrees with the visual measuring performed during restoration of the stilling basin.

Localised erosion should not be admitted downstream from a spillway with a hydraulic jump stilling basin whose dimensions have been determined using classical criteria. Nevertheless, this is never accomplished. In fact, erosion downstream from a stilling basin whose length is equal to the theoretical jump's length is never zero.

It is usually assumed that the velocity profile in the downstream section is uniform (channel turbulent flow profile), but this is not so because the Froude Number at Section 1 decreases. This shows that the speed on the surface is very low and near the bottom is high and that the eroding capacity is much higher at average speed. Some kind of non-eroded basalt has been found where large flatland dams have been built. Stilling basins have been designed using standard criteria with a length representing 60% of the theoretical length. When end sills are high enough they generate forced hydraulic jump conditions that favour macroturbulence and riverbed erosion.

A warning against considering that a rocky bed submitted to severe pressure fluctuations behave as an alluvial riverbed is given. Equilibrium is dynamical in this last case, reaching a final depth condition after scouring On the other hand, when the bed is composed of meteorized rock with blocks of different dimensions, removal of one of the blocks is irreversible. The eroding process on rocky beds is completely different from granular material erosion in the laboratory.

The presence of an end sill is beneficial when the riverbed is alluvial. When riverbed is composed by rock, end sills favour local scour generating an increase of speed fluctuations, concentrating energy around a dominant frequency that trends to remove large blocks.

4. UNCONVENTIONAL CASE STUDY

The Río Hondo Dam is located on the Dulce River, north of Argentina, and is used for flood attenuation, irrigation and hydroelectric power supply.

The discharge structures of the dam are made up of two outlets and four valves for water diversion. The spillway has a free L = 151 m length and a design head H = 3m, that allows for a maximum spillway discharge of 1,525 m^3/s. The maximum discharge flowing through all outlet devices is Q = 2,248 m^3/s. Spillway and outlet structures discharge into a horizontal stilling basin 80 m long, with a +245.92 topographic level, an upstream width of 166.5 m and a downstream width of 182.5 m. The dam has a line of baffle piers and a stepped end sill, with a +249 m topographic top level (Figure 3).

Figure 3: Río Hondo Dam stilling basin

The dam began operations in the sixties. The "wall" produced an abrupt discontinuity in sediment transport in the river. A typical process of scour downstream dams in fluvial streams developed along several kilometers as a progressive and permanent phenomena, searching in an asymptotic form a new riverbed profile. Near the stilling basin the bed level decreased from +250 m to less than + 247 m.

This process modified hydrodynamic conditions downstream from the stilling basin and exerted a greater impact on the riverbed than designers expected. The hydraulic jump cannot be placed inside the concrete structure and a jet jump from the end sill to the riverbed changed local scour conditions.

The authors have presented an unconventional methodology based on the following concept: even if the structure is a stilling basin, with baffle piers for forced energy dissipation, the critical flow over the end sill generates a new hydrodynamic condition, and the accelerated flow acts as a submerged jet, like the hydrodynamic configuration of a ski-jump spillway, but with a very low hydraulic head.

The proposed maximum depth methodology for scour estimation is based on INCYTH's equation, which states that the "ski-jump floor" level is given by the top of the end sill level and the total head (Hups – Hdown) must to be calculated as:

$$\Delta H = h_d + V_d^2/(2\ g).$$

This equation takes into account friction energy losses along the flow through the spillway and stilling basin and local energy losses due to baffle piers and the end sill.

The maximum depth methodology for scour estimation recorded a maximum flood of y = 5.26 m, very close to the prototype measured value.

This methodology is used when sediment discontinuity decreases tailwater level and a critical flow is produced at the end of the structure. This abnormal situation is frequent in small dams in the northwestern region of Argentina.

5. CONCLUSIONS

Local scour downstream from ski-jump spillways can be estimated by means of an oversimplified equation, without knowledge of the fractured rock size. The equation is sufficiently good for a first value of maximum depth of scour at long period. A physical model with movable bed makes it possible to assess the problem. On the other hand, the hydraulic model is always required to determine the recycling currents originated in the impact jet zone.

The phenomenon of local erosion downstream from hydraulic jump energy dissipators in large flatland dams, where the riverbed is made up of rocks that have undergone severe pressure fluctuations, is analyzed using the experimental results of pressure fluctuation amplitudes downstream from a forced jump with a two-dimensional barrier (end sill).

Due to the progressive decrease of the river bottom downstream from the Río Hondo Dam that reservoir sediments have produced for more than forty years, the tailwater level is now below the sequent depth and the hydraulic jump cannot be formed in the stilling basin. The flow downstream from the stilling basin is accelerated so as to increase local scour. The structure acts as a ski-jump but with strong energy losses along the spillway and stilling basin (due to baffle piers, friction losses and end sill).

The authors have presented an unconventional methodology based on the following concept: even if the structure is a stilling basin, with baffle piers for forced energy dissipation, the critical flow over the end sill generates a new hydrodynamic condition, and the accelerated flow acts as a submerged jet, like the hydrodynamic configuration of a ski-jump spillway, but with a very low hydraulic head.

6. REFERENCES

Balloffet, A. (1987): "Erosión y sedimentación: revista de métodos de análisis" (Scour and sedimentation: revue of methods for analysis), *First Argentine Seminar on Large Dams*, Ituzaingó, Corrientes, Vol. 2, pp. 151-189.

Chividini, M.F. et Al (1983): "Evaluación de la socavación máxima aguas debajo de aliviaderos en salto de esquí" (Evaluation of maximum scour downstram ski-jump spillways), *Proceedings of the XI Water National Congress*, Córdoba, Argentina, Vol. 6, pp. 187-210.

Keming, A. et Al (1987): "Free jet scour to rock riverbed", *XXII IAHR Congress*, Energy dissipation Seminar, Lausanne, Switzerland.

Lopardo, R.A. et Al (1987): "The role of pressure fluctuations in the design of hydraulic structures" in *Design of Hydraulic Structures*, edited by R. Kia and M.L. Albertson, Colorado State University, Fort Collins, pp. 161-175.

Lopardo, R.A. & Lapetina, M. (1997): "Local scour on rocky beds downstream stilling basins, in *Management of Landscapes Disturbed by Channel Incision*, edited by S.S.Y. Wang, E.J. Langendoen and F.D. Shields, The University of Mississippi, Oxford, USA, pp. 288-295.

Lopardo, R. & Sly, E. (1992): "Constatación de la profundidad máxima de erosión aguas debajo de aliviaderos en salto de esquí" (Validation of the maximum depth of scour downstream ski-jump spillways), *Revista Latinoamericana de Hidráulica IAHR*, Sao Paulo, Brasil, N° 4, pp. 7-23.

Mason, P.J. & Arumugam, K. (1985): "Free jet scour below dams and flip buckets", *ASCE, Journal of Hydraulic Engineering*, Vol. 11, N° 2, pp. 220-235.

Oliveira Lemos, F. & Matías Ramos, C. (1984): "Hydraulic modelling of free jet energy dissipation", *Symposium on Scale Effects in Modelling of Hydraulic Structures*, Esslingen am Neckar, Germany, pp. 7.6/1-7.6/5.

Riedel, R. (1989): "Socavación aguas abajo del salto de esquí del vertedero de la presa de Colbún" (Scour downstream the ski-jump spillway of Colbún Dam), *IX National Congress of Chilean Hydraulic Engineering Society*, Santiago, Chile.

Rock Scour due to falling High-velocity Jets - Schleiss & Bollaert (eds)
© 2002 Swets & Zeitlinger, Lisse, ISBN 90 5809 518 5

Quantification of extent of scour using the Erodibility Index Method

G.W. Annandale
Engineering & Hydrosystems Inc., Highlands Ranch, Colorado, USA

ABSTRACT: The paper presents an overview of applying the Erodibility Index Method to predict extent of scour. With the variation of the erosive power of water as a function of space and time known for the flow conditions under consideration it is possible to predict scour extent for any flow condition and any earth material by using this method. Methods to determine spatial and temporal variation in the erosive power of water can be determined by developing equations from basic principles of hydraulics, empirical methods, physical hydraulic model studies or computer simulation. This paper presents two examples that use empirical relationships determined by physical hydraulic model studies. The one example relates to scour around bridge piers and the other to plunge pool scour. A summary of the Erodibility Index Method that is used in this approach is provided in a companion paper to this volume (Annandale 2002a) as is the method that is used to quantify the relative ability of rock to resist scour (Annandale 2002c).

1 INTRODUCTION

The approach that is used to predict the extent of scour using the Erodibility Index Method is presented. Methodologies for calculating bridge pier and plunge pool scour extents, and results of near-prototype tests pertaining to scour of simulated rock are presented. Variation of the erosive power of water as a function of scour depth is defined by means of empirical relationships that were developed with the results from physical hydraulic model studies. The relative ability of earth material to resist scour is quantified by means of the Erodibility Index.

2 BRIDGE PIER SCOUR

The first bridge pier scour analysis using the Erodibility Index Method was conducted for the Northumberland Strait Bridge (Anglio et al. 1996) (Figure 1). This analysis entailed verification of the Erodibility Index Method by using on site material properties, estimates of the erosive power of water and observed scour of rock around one of the bridge piers. In order to analyze the scour conditions at the base of the remainder of the bridge piers laboratory studies were conducted to quantify the relative magnitude of the erosive power of water around the bridge piers for varying flow conditions. The verified relationship and estimates of the relative magnitude of the erosive power of water for design conditions were used to predict the likelihood of scour at other bridge piers and to design countermeasures.

Figure 1. Northumberland Strait Bridge to Prince Edward Island, Canada.

The scour analysis that was conducted for the new $2.54 billion Woodrow Wilson Bridge that will be built over the Potomac River close to Washington D.C. (Figure 2) is briefly summarized in what follows to illustrate application of the Erodibility Index Method to bridge pier scour.

2.1 Erosive Power of Water

The erosive power of water changes as a function of scour depth. It is at its highest just before scour commences at the base of a bridge pier, steadily decreasing as the scour-hole increases in depth. Before scour commences the turbulence at the base of the bridge pier is at its highest, as is the erosive power of the water. As the scour increases in depth, streamlined flow conditions develop within the scour hole, leading to decreased turbulence intensity. As the turbulence intensity within the scour hole decreases so does the erosive power of water.

Dimensionless relationships between scour depth and the erosive power of water were developed for rectangular and round bridge piers (Smith et al., 1997). The relationships for round and rectangular piers (parallel to the flow) are shown in Figure 3. The ordinate scales the ratio P/P_a and the abscissa the ratio y_s/y_{max}, where P = stream power at base of the bridge pier;

Figure 2. Existing Woodrow Wilson Bridge that will be replaced with a new bridge on the same site.

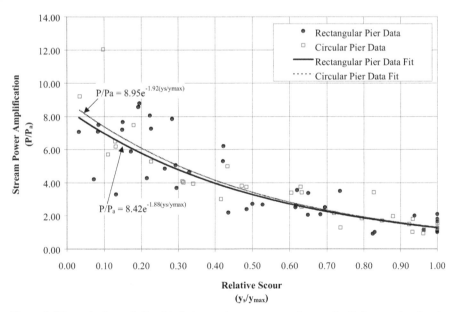

Figure 3. Dimensionless relationship between stream power and scour depth for round and rectangular piers.

P_a = stream power in the approach flow upstream of the bridge; y_s = scour depth; y_{max} = maximum scour depth.

Quantification of the axes in Figure 3 requires estimates of the approach stream power (P_a) and the maximum scour depth (y_{max}). The equation that is used to calculate the approach stream power is based on,

65

$$P_a = \tau v \qquad\qquad (1)$$

where P_a = approach stream power per unit area of the bed (W/m^2), τ = shear stress on the bed (N/m^2) and v = velocity (m/s).

By writing shear stress τ as a function of the unit weight of water, flow depth and energy slope, approach stream power can also be expressed as:

$$P_a = \gamma\, d\, s\, v \qquad\qquad (2)$$

where d = flow depth (m), γ = unit weight of water (N/m^3), s = dimensionless energy slope (or bed slope in the case of uniform, steady flow), and v = velocity (m/s).

An estimate of y_{max} can be obtained by making use of the bridge pier scour equation in HEC-18 (FHWA 1995), which is based on an envelope curve embracing a large number of bridge pier scour experiments. This equation (presented below) is considered to provide a conservative estimate of maximum scour depth in non-cohesive granular earth material.

$$\frac{y_s}{y_1} = 2.0 \cdot K_1 \cdot K_2 \cdot K_3 \cdot \left(\frac{a}{y_1}\right)^{0.65} Fr_1^{0.43} \qquad\qquad (3)$$

where y_s = scour depth (ft), y_1 = flow depth directly upstream of the pier (ft), K_1 = correction factor for pier nose shape, K_2 = correction factor for angle of attack of flow, K_3 = correction factor for bed condition, a = pier width (ft), L = length of pier (ft), Fr_1 = Froude Number = $V_1/(gy_1)^{1/2}$, and V_1 = mean velocity of flow directly upstream of the pier (ft/s).

With y_{max} assumed to be the maximum scour, the scour depth estimated with the Erodibility Index Method can never exceed this value. The reason for this is that the equation has been developed for a weak earth material (non-cohesive granular soils), and that the scour depth for more resistant earth material can reasonably be expected to be less. The range of scour depth estimates for this method is therefore $0 \le y_s \le y_{max}$.

The two curves on the graph can also be represented by the following two equations:

Rectangular Piers: $P/P_a = 8.42 \cdot e^{-1.88 \cdot (y_s/y_{max})}$ $\qquad\qquad$ (4)

Circular Dolphins: $P/P_a = 8.95 \cdot e^{-1.92 \cdot (y_s/y_{max})}$ $\qquad\qquad$ (5)

Quantification of these equations were used to calculate the change in the erosive power of water for the new Woodrow Wilson Bridge for the 100- and 500-year floods (Figure 4).

2.2 Earth Material Ability to Resist Scour

By making use of geologic cores and borehole data it was possible to quantify the relative ability of the earth material underlying the bridge to resist erosion (Annandale 2002c). The variation of scour resistance as a function of elevation for the Woodrow Wilson Bridge is shown in Figure 4.

2.3 Extent of Scour

The extent of scour that is anticipated for the new Woodrow Wilson Bridge is shown in Figure 4, which shows the change in the erosive power of water as a function of potential scour depth for the 100- and 500-year floods, as well as the relative ability of the earth material to resist erosion.

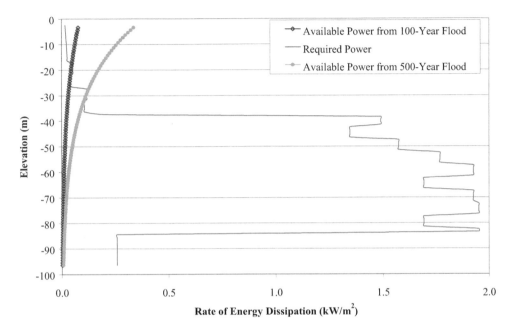

Figure 4. Comparison between available stream power for the 100- and 500-year floods, and the stream power required to cause scour in the underlying earth material.

The figure shows that the maximum extent of scour is unlikely to ever exceed about -38m elevation. The material at this elevation is so strong that it is unlikely that erosive power resulting from the presence of the bridge pier will ever be greater than the required stream power at this elevation. Whenever the required stream power exceeds the available stream power, scour is unlikely to occur.

3 PLUNGE POOL SCOUR

The Erodibility Index Method was successfully used to analyze plunge pool scour at Bartlett Dam, Arizona (Cohen and von Thun 1994) and erosion thresholds on the rock abutments of Gibson Dam, Montana (Annandale et al. 2000) (Figure 5). The plunge pool downstream of Bartlett Dam experienced approximately 30m of scour in granite, whereas Gibson Dam only experienced minor damage to its abutments during an extreme flood that overtopped the dam.

The example pertaining to plunge pool scour presented in this paper represents scour that was measured under controlled conditions during near-prototype tests. The scour was measured in both non-cohesive granular material and simulated rock. The results are presented here to demonstrate how the method to calculate scour extents is applied.

67

Figure 5. The Erodibility Index Method was successfully used to analyze plunge pool scour at Bartlett Dam, Arizona (left) and erosion thresholds at Gibson Dam, Montana (right).

3.1 Experimental Facility

A prototype experimental facility located at Colorado State University in Fort Collins, Colorado, USA was used to test the Erodibility Index Method. The facility includes a basin, 10 m wide by 16.75 m long and 4.5 m deep, and a 8.7 cm by 3.05 m wide nozzle discharging up to 3.4 m³/s at angles ranging from zero to forty-five degrees from vertical (Figure 6). Two sets of experiments were conducted, one testing the response of non-cohesive granular earth material and the other testing the response of simulated rock to plunging jets.

A small-scale facility that was used to develop empirical equations for calculating the variation in velocity resulting from aerated plunging jets in plunge pools was also built. The equations developed from this study are used to calculate the extent of scour in plunge pools when using the Erodibility Index Method. Scour depth is determined by comparing available and required stream power (Annandale 2002a), as illustrated in what follows.

3.2 Available Power

The power (kW/m²) available to erode material is a function of the jet hydraulics. From Bohrer & Abt (1996), the velocity along the centerline of a jet in a plunge pool is a function of the velocity at impact, the angle of impact, the air concentration of the jet at impact, given by the ratio of air and water densities, and gravitational acceleration. Equation 6 describes this functional relationship, followed by the limits of application. Equation 7 yields the distance along the centerline.

$$-\ln\left(\frac{v}{V_i}\right) = -0.5812\ln\left[\left(\frac{\rho_i}{\rho_w}\right)\left(\frac{V_i^2}{gL}\right)\right] + 2.107 \tag{6}$$

Figure 6. Experimental facility for testing response of non-cohesive granular earth material and simulated rock to plunging jets, showing simulated rock response only.

$$-0.29 < \ln\left[\left(\frac{\rho_i}{\rho_w}\right)\left(\frac{v_i^2}{gL}\right)\right] < 2.6$$

$$L = \frac{z_j - z_{j+1}}{\cos\alpha} \tag{7}$$

where v = velocity at distance L along jet centerline beneath water surface (m/s); V_i = velocity at location where jet impacts at water surface (m/s); ρ_i = mass density of aerated jet at impact with water surface (kg/m³); ρ_w = mass density of water (kg/m³).

The rate of energy dissipation in the plunge pool (power available for scour) can be expressed as a discretized function of the total head at various elevations along the centerline of the submerged jet. Equation 8 shows a discrete calculation for the change in energy, ΔE_j, between points j and j+1. As the velocity decays, with decreasing elevation, or increasing displacement along the jet centerline, the total head decreases. Equation 9 yields the corresponding available power, p_{Aj}.

$$\Delta E_j = \frac{v_j^2 - v_{j+1}^2}{2g} + \frac{P_j - P_{j+1}}{\gamma} + z_j - z_{j+1} \tag{8}$$

69

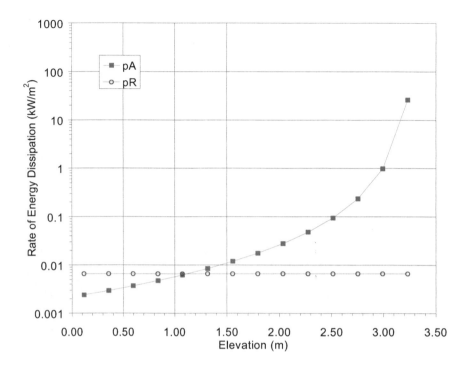

Figure 7. Power available and power required.

$$\frac{dp_{A_j}}{dz} = \frac{\gamma v_j \Delta E_j}{1000(z_j - z_{j+1})} \tag{9}$$

3.3 Scour calculation

The scour depth is calculated by comparing the stream power that is available to cause scour and the stream power that is required to cause scour. Figure 7 shows an example of one of the comparisons for calculation of scour in granular earth material. The stream power required to cause scour in this case is constant, resulting from the homogeneity of the earth material.

3.4 Granular Material: Comparison with Observed Scour

Figure 8 compares calculated and observed scour elevations for twelve experiments. On average, the predicted scour depths are approximately equal to the observed scour depths. The procedure for estimating the depth of scour in a plunge pool accounts for angle of jet impact, jet aeration and material properties.

Figure 9 shows the comparison between observed and calculated scour using equations of Mason & Arumugamand (1985) and Yildiz &Üzücek (1994) for the same data set.

3.5 Simulated Rock: Comparison with Observed Scour

Table 1 compares scour that was observed for simulated rock in the experimental facility and those calculated using Mason & Arumugam (1985) and using the Erodibility Index Method.

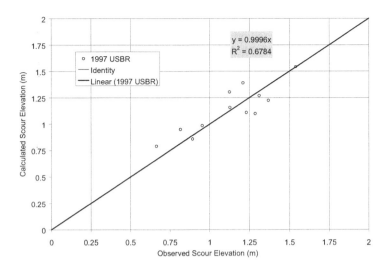

Figure 8. Comparison of observed and calculated scour in granular material.

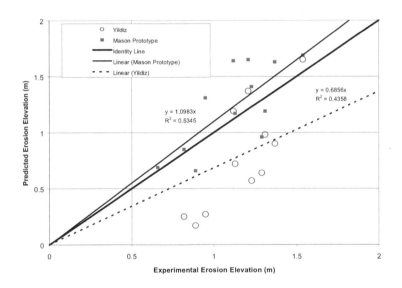

Figure 9. Comparison between observed scour data and calculated data using Mason & Arumugamand (1985) and Yildiz &Üzücek (1994).

Table 1. Comparison of experimental data with Mason equation.

	1.133 m^3/s	1.274 m^3/s	1.416 m^3/s
Depth of Scour - Experiment	0.67 m	0.70 m	0.73 m
Depth of Scour – Erodibility Index	0.67 m	0.67 m	0.67 m
Depth of Scour - Mason	0.95 m	1.01 m	1.08 m

71

4 SUMMARY

The paper summarizes application of the Erodibility Index Method to calculate the extent of scour. It illustrates that the erosion threshold defined by the Erodibility Index Method (Annandale 1995 and Annandale 2002a) can be used to estimate the erodibility of any earth material, and that the method can be used to estimate the extent of scour for any flow condition, provided that variation in the erosive power of water, expressed in terms of the rate of energy dissipation, can be quantified as a function of space and time. The latter can be accomplished by means of empirical equations, hydraulic equations developed from basic principles of hydraulics, physical model studies, and computer simulation. Comparison between observed and calculated scour is satisfactory for purposes of practical application.

5 REFERENCES

Annandale, G.W. 2002a. The Erodibility Index Method: An Overview, *International Workshop on Rock Scour*, EPFL, Lausanne, Switzerland, September 25-28, 2002.

Annandale, G.W. 2002b. Quantification of the Relative Ability of Rock to Resist Scour, *International Workshop on Rock Scour*, EPFL, Lausanne, Switzerland, September 25-28, 2002.

Annandale, G.W., Wittler, R. and Scott, G.A. 2000. Scour Downstream of Dams, *Scour Symposium*, ISSMGE GeoEng2000 Conference, Melbourne, Australia.

Annandale, G.W. 1995. Erodibility, *Journal of Hydraulic Research*, Vol. 33, No. 4, pp. 471-494.

Anglio, C.D., Nairn, R.B., Cornett, A.M., Dunaszegi, L., Turnham, J. and Annandale, G.W. 1996. Bridge Pier Scour Assessment for the Northumberland Strait Crossing, Coastal Engineering 1996, *Proceedings of the Twenty-fifth International Conference held in Orlando*, Florida, September 2-6, 1996, Billy L. Edge, Editor.

Bohrer, J.G. and Abt, S.R. 1996. Plunge Pool Velocity Prediction of Rectangular, Free Falling Jets, Dam *Foundation Erosion Study, Phase II report*, US Bureau of Reclamation, Denver, Colorado.

Cohen, E. and Von Thun, J. L., 1994, Dam Safety Assessment of the Erosion Potential of the Service Spillway at Bartlett Dam, *Proc. International Commission on Large Dams*, Durban, South Africa, pp. 1365-1378.

Federal Highway Administration. 1993. HEC-18, *Evaluating Scour at Bridges*, Edition 2, FHWA# IP-90-016., Washington D.C.

Mason, P.J., Arumugam, K., Free Jet Scour Below Dams and Flip Buckets, *Journal of Hydraulic Engineering*, Vol. 111, No. 2, ASCE, February 1985.

Smith, S. P., Annandale, G. W., Johnson, P. A., Jones, J. S. and Umbrell, E. R. (1997), Pier Scour in Resistant Material: Current Research on Erosive Power, in *Proc. Managing Water: Coping with Scarcity and Abundance, 27th Congress of the International Association of Hydraulic Research*, San Francisco, California, pp. 160-165.

Yildiz, D., Üzücek, E., "Prediction of Scour Depth From Free Falling Flip Bucket Jets." *Intl. Water Power and Dam Construction*, November, 1994.

Rock Scour due to falling High-velocity Jets - Schleiss & Bollaert (eds)
© 2002 Swets & Zeitlinger, Lisse, ISBN 90 5809 518 5

A review on physical models of scour holes below large dams in Iran

J. Attari
Assistant professor, water department, Power & Water Institute of Technology, Tehran, Iran

F. Arefi
Head of hydraulic section, dams and power plant dept., Mahab Ghods Cons. Eng. Co., Tehran, Iran

F. Golzari
Senior research officer, hydraulic structures section, Water Research Centre, Tehran, Iran

ABSTRACT: Water nappes or plunging jets, released by spillways terminating to flip buckets, create scour holes at downstream valleys of large dams. Physical modelling and empirical formulae are used to predict geometry of the scour holes. In this paper, four physical models of unlined plunge pools below large dams in Iran were reviewed. Results showed that the models generally overestimated the ultimate scour depth in comparison with the selected empirical formulae. However, good agreement of the observed scour depths at Karun 1 prototype was found with the formulae. Furthermore, the longitudinal location of the scour hole was investigated and compared with semi-empirical relationships for jet trajectory.

1. INTRODUCTION

One of the most effective and economical ways of releasing excess water from dam reservoirs is to discharge the flood as a free jet to the downstream valley. Free overfalls, ski jumps, orifices and other types of spillways terminating to flip buckets have been widely used for this purpose in different large dams (e.g. ICOLD 1987). However, the high velocity jets, issued from such spillways, will cause excessive scouring at the downstream valley. This may endanger the dam safety (e.g. Kariba and Kaborabosa dams in South Africa) unless the energy of the flow has been dissipated enough in a plunge pool (Vischer & Hager 1998). Such a pool may be created naturally by the plunging jet or can be pre-excavated. The latter may be un-lined or in some cases might have a concrete lining or slab (Ervine et al. 1997).

The first step in design of plunge pools is to predict the scour hole geometry (Whittaker & Schleiss 1984, Attari 1993). This includes: determination of the jet impact location, estimation of the ultimate depth of scouring, the side slopes and the lateral scour development. Currently, empirical formulae developed from previous physical model studies and limited prototype observations are used for prediction of ultimate scour depth for preliminary designs. Owing to significant difference between the values predicted by various formulae and other uncertainties, physical modelling is usually carried out at detailed design stage, to foresee the entire geometry of the scour hole.

There are many large dam projects under design and construction in Iran (Hydropower & Dams 1999). Considering large floods and high heads of these dams, spillways terminating to flip buckets and plunge pool type energy dissipaters are widely used. The objective of this paper is to report the experience with prediction and modelling of scour holes used for design of plunge pools below large dams in Iran.

2. MAIN FEATURES OF THE DAM PROJECTS

The main features of the selected dam projects in Iran, referred in this paper, are summarised in Table 1.These are all high head dams with large flood discharge. In all of these projects, spillway terminates into a flip bucket structure.

Table 1.Main features of the selected projects

Name of Dam	Type of Dam	Height of dam (m)	Discharge capacity (m³/s)	Spillway type	Chute width (m)	Year of completion
Karun 1	Arch	200	16200	chute	54.5	1976
Marun	Rockfill	165	11330	chute	62.2	2001
Karun 3	Arch	205	21000	chute+orifice+ overfall	49.5	2003
Agh Chay	Earthfill	110	3680	chute	80	2005

The chute spillway of Karun 1 dam was completed in 1976 but suffered from severe cavitation damage that was repaired annually. In the process of installation of aerators on the chute spillway in 1993, an extreme flood destroyed the old elevated deflected flip bucket. During rehabilitation of the spillway in 1993-4, aerators were installed at the spillway and a new straight bucket was constructed at an upper elevation (Jalal Zadeh et al. 1994). Since then the spillway has operated for a few times and has performed favourably.

3. PHYSICAL MODELS DESCRIPTION

Specifications of physical models for the four projects are given in Table 2.All of the models were scaled according to the Froude law of similarity. In this paper, the Karun1 model refers to the very last physical model with the straight bucket. Two types of materials (i.e. cohesive and non-cohesive) were used in physical models of the Karun 3 dam built at Water Research in Tehran and Acres laboratory in Niagara falls. Among different combination of cohesive materials tested in Acres laboratory, only the final combination is given in Table 2.

Table 2. Specifications of the models

Model	Scale	Material type	Material combination(%)	Reference
Marun	1:76.5	Non-cohesive	gravel d_{50}=12.5mm	WRC (1986), Mahab Ghods(1996)
Karun 1	1:62.5	Non-cohesive	gravel d_{50}=9.5mm	WRC (1993), Mahab Ghods(1994)
Agh Chay	1:40	Non-cohesive	gravel d_{50}=6 mm	Golzari(2001), WRC (2002)
Karun 3 (a)	1:80	Non-cohesive	gravel	WRC (1995)
Karun 3 (b)	1:80	Cohesive	gravel(45),sand(30),gypsum (12.5),water(11.5),cement(1)	
Karun 3 (c)	1:90	Non-cohesive	gravel d_{50}=5.6mm	Mahab Ghods Acres general partnership (1993)
Karun 3 (d)	1:90	Cohesive	gravel(42.8),sand(20),water (19.7), grit(10), bentonite(7.5)	

4. ULTIMATE DEPTH OF SCOURING

The most important parameter, which defines the scour hole geometry (Fig.1), is the ultimate depth of scouring ($Y = z_{sc}+h_2$). This depth was calculated using various empirical formulae, mostly reviewed in Whittaker & Schleiss 1984, at preliminary design stage. Among these the famous relationships listed in Table 3 were found more appropriate.

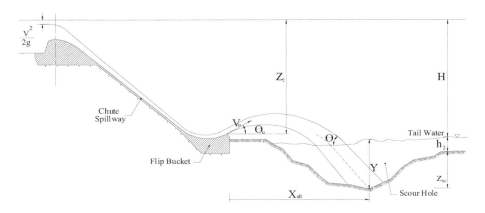

Figure 1. Definition sketch of a scour hole

Table 3. Some famous empirical formulae for prediction of ultimate scour depth

Name of formula	Year	Equation	Reference
Veronese	1937	$Y=1.9q^{0.54}H^{0.225}$	ICOLD (1987)
Modified Veronese	1994	$Y=1.9q^{0.54}H^{0.225}\sin\theta$	Yildiz & Uzucek (1994b)
Martins	1975	$Y=1.5q^{0.6}Z_2^{0.1}$	Martins (1975)
Chian Min Wu	1973	$Y=1.18q^{0.51}H^{0.235}$	Whittaker & Schleiss (1984)
Mason	1985	$Y=3.27q^{0.6h}H^{0.15}h_2^{0.05}g^{-0.3}d^{-0.1}$	Mason & Arumugam (1985)

4.1 Present model results

The results for the ultimate depth of scouring for the mentioned projects are given in Figure 2 and compared with the selected famous empirical relationships listed in Table 3. Application of the Mason (1985) formula, specifically given for model data, dramatically overestimated the present model results. This formula seems to be out of range of applicability for present study owing to high values of the unit discharge and heads. The present measured depths of scours are significantly less than the values predicted by the Veronese formula. This verifies the general belief that the Veronese formula provides an upper limit (USBR 1977).

Figure 1a compares the data with the Martin formula recommended by the ICOLD (1987). This relationship incorporates head Z_2 (i.e. difference between levels of water in reservoir and flip bucket), which is independent of pool water depth. However, it still provides a satisfactory estimation of the ultimate scour depth.

Figure 1b demonstrates that the Chian Min Wu relationship predicts the scour depth close to the Martin formula and has the advantage of taking tailwater into account.

Figure 1c shows that the measured values in the present physical models are well correlated around the Modified Veronese formula. This verifies that incorporating the jet impact angle (θ) improves the scour depth relationships as demonstrated for three prototypes in Turkey (Yildiz & Uzucek 1994b).

Figure 2 shows that the results for the non-cohesive Karun3 models (a, c) were near each other. Despite the different percentage of the cohesive materials (Table 2) used for the Karun 3 models (b) and (d), roughly similar results were achieved (Fig.2). The results of the non-cohesive models (a, c) were not significantly higher than their cohesive counterparts (b, d). This could be attributed to effect of sidewalls of the narrow Karun 3 valley, which were modelled rigidly in both cases and thus prevented lateral scour in the models. Otherwise a deeper scour hole would have been expected by using cohesive materials for modelling (Whittaker & Schleiss 1984).

Physical modelling of scour holes on rocky bed rivers involves various sources of uncertainties and therefore might not accurately determine the ultimate depth of scouring, however they are useful tools for prediction of overall scour hole geometry.

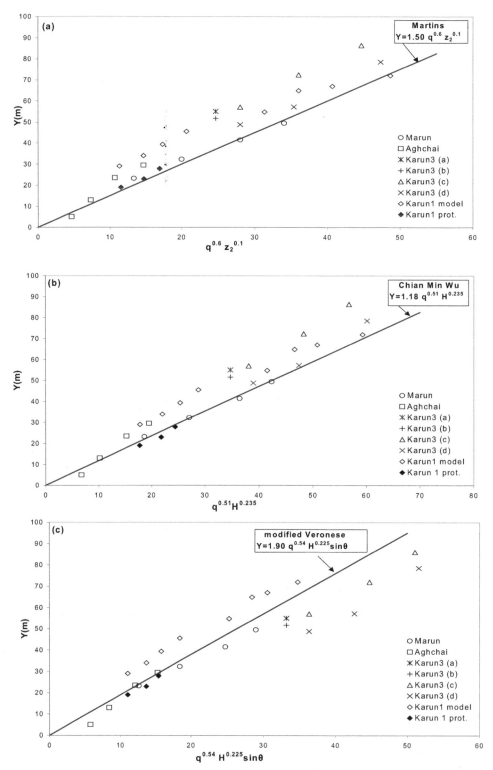

Figure2. Comparison of the measured ultimate scour depths with empirical relationships

4.2 Karun 1 prototype scour observation

Two bathymetric surveys were carried out downstream of the Karun 1 dam since rehabilitation of its spillway. The first one was in 1994 after trial operation of spillway with discharges of about 500 and 750 m³/s released from one bay of the chute spillway (Mahab Ghods 1994). The 2nd survey, made in 2000, shows the scour created by the flood with a peak discharge of 2320 m³/s released from all three bays (Fig.3). From these surveys, the observed scour depths for the mentioned flow conditions were plotted on Figure 1.This graph shows that the observed depths of scour in Karun 1 prototype were less than their corresponding values in model. This agrees with the model-prototype measurements for Itaipu dam (Torales et al. 1994). Such a behavior can generally be expected owing to scale effects and phenomena such as jet disintegration and air entrainment that reduces the actual depth of scour but cannot be reproduced in scale models. It is interesting that the observed scour depth in Karun 1 is near the values predicted by the selected empirical relationships (Fig.2). Close agreement of the scour depths predicted by these relationships (i.e. Martin and Chian Min Wu) with actual values were also found for three dams in Turkey (Yildiz & Uzucek 1994a). However, in the absence of a large number of prototype data, it is not possible to accurately predict the scour depth by simple empirical relationships.

5. LONGITUDINAL LOCATION

In design of spillways terminating to flip buckets, it is important to estimate the length of the jet trajectory that determines the location of the scour hole. For this purpose, formulae based on the theoretical profile of free jets incorporating some empirical correction factors to allow for air resistance are usually used. But such an analytic method cannot yet accurately predict the jet trajectory owing to very limited prototype data and difficulties associated with such field measurements. Alternatively, physical models are still used for the estimation of the jet trajectories but air resistance jet disintegration and air entrainment cannot be simulated in the scale models.

To estimate jet trajectory lengths, Rae & Castro (1994) examined surveyed cross sections of seven prototype scour holes including the old Karun 1 spillway. They compared this information with the USBR (1977) and Kawakami (1973) relationships and concluded that Kawakami equation generally provided better estimates of trajectory than the USBR equation.

Figure 3 shows the surveys of the scour holes of present study. From these measurements, average longitudinal distances from end of flip buckets to the deepest scour locations, x_{ult}, were extracted (Table 4). The results are compared with the Kawakami (1973) and the USBR (1977) formulae for the jet trajectory (Table 4).

Table 4.Comparison of the longitudinal distance from end of flip bucket to the deepest scour hole location

Name of dam	Unit disc. $q(m^2/s)$	Jet velocity V_0 (m/s)	Bucket lip angle (θ_0)	Jet length x_{ult}(m)			
				Calculated		Measured	
				USBR	Kawakami	Model	Comment
Marun	159	50.6	40	241	140	264	extrapolated
Agh Chay	46	31.6	40	131	111	120	
Karun 3	297	40.6	30	239	156	227	cohesive
Karun1 (new bucket)	40	48.3	20	241	143		165m (prototype)

Table 4 shows that the results obtained in model are close to the USBR formula that is essentially the theoretical projectile profile incorporating a reduction factor of 0.9.This confirms the fact that the air entrainment and jet disintegration is not reproduced in physical models.

The result obtained from a field observation in Karun 1 prototype is only about 10% higher than the jet length predicted by the Kawakami relationship whereas it is significantly lower than the value from the USBR formula (Table 4). This agrees with the above-mentioned conclusion of Rae & Castro (1994).

Although more prototype data are still required to improve the Kawakami correction factor such as depth of water at the flip bucket and the initial turbulence level of the jet, it still provides a good basis for estimation of the jet trajectory in atmosphere (Locher & Hsu 1984).

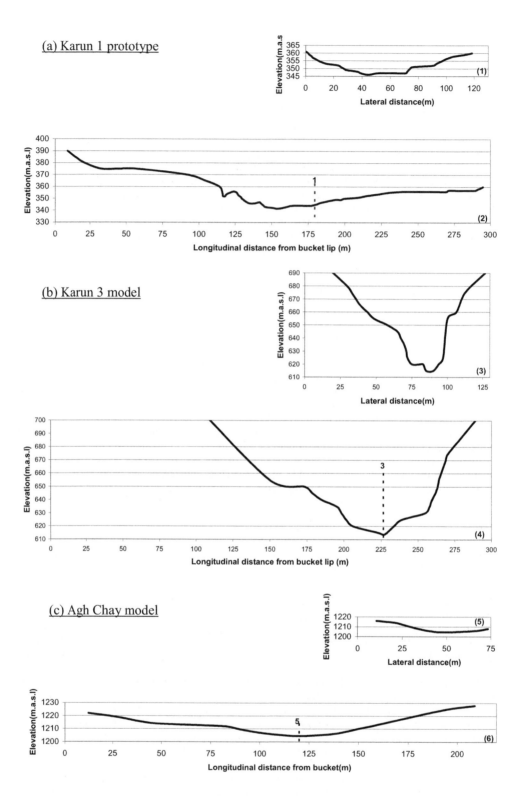

Figure3. Survey of scour holes downstream of the dam projects in Iran

78

6. CONCLUSIONS

Scour on rocky bed rivers is a complex problem. Review of Karun 1 prototype and the four physical models of scour holes, created by flip buckets below large dams in Iran, showed that:

- The physical models in the present study overestimated the ultimate scour depth. However they were useful tools for prediction of overall scour hole geometry.
- Accurate prediction of scour depth is not possible by empirical formula. However, three formulae (i.e. Martins, Chian Min Wu, Modified Veronese) were found appropriate for estimation of scour depths in trajectory basins at preliminary design stage.
- A good agreement of the observed scour depth in Karun 1 prototype was found with the mentioned empirical formulae and similar dams in Turkey.
- The Kawakami formula was found appropriate for prediction of scour hole location at design stage.
- There is a need for more investigations, especially prototype observations, to refine the empirical formulae for prediction of scour hole geometry.

ACKNOWLEDGEMENTS

The authors acknowledge the permission of the Water Research Centre and the Mahab Ghods consulting Eng. Co. to use their reports for preparation of this review article. The clients for the projects were Iran Water and Power Development Co., Khozestan Regional Power and Water Authority and West Azarbayjan Regional Water Authority. The authors are grateful to the above-mentioned companies, which are all affiliated to the Ministry of Energy of I.R. Iran. Some of the tests for the Karun 3 project were carried out by the Acres Company laboratory in Niagara falls that is gratefully acknowledged.

REFERENCES

Attari, J. 1993. Plunge pools downstream of free overfall spillways: A review report. Tarbiat Modares University. Tehran. (In Farsi).

Ervine, A. et al. 1997. Pressure fluctuations on plunge pool floors. *J. of Hydraulic Research IAHR,35(2).*

Golzari, F. 2001. Physical model study of flip buckets hydraulics in large dams. *Proc. Int. conf. of hydraulics structures.* Kerman . Iran . (In Farsi).

Hydropower & Dams. 1999. World Atlas and industry guide. *Aqua Media International.*

ICOLD.1987. Spillway for dams. Bulletin 58.

Jalalzadeh, A. et al. 1994. Spillway damage and the construction of a new flip bucket at Karun 1. *Int. Journal of Hydro Power & dams, Nov.*

Kawakami, K. 1973.A study on the computation of horizontal distance of jet issued from ski-jump spillway. *Transaction of the Japan Society of Civil Engineers, Vol. 5.* Cited in P. Novak (ed.) Developments in hydraulic engineering Vol.2.

Locher, F & E.Y.Hsu.1984. Energy dissipation at high head dams. In P. Novak (ed.), *Developments in hydraulic engineering Vol.2.*

Mahab Ghods Cons. Eng. Co.1996.Report on revision of spillway and aerator design of Marun dam. (In Farsi).

Mahab Ghods Cons. Eng. Co.1994.Turn key project for rehabilitation and repair of Karun 1 dam: Vol.4: Hydraulic design report. (In Farsi).

Mahab Ghods Acres general partnership.1993.Karun 3 development project: Complementary comprehensive hydraulic model study report. Niagara falls. Ontario.

Martins, R. 1975. Scouring of rocky river beds by free jet spillways. *Water power & dam construction, April.*

Mason, P.J. & Arumugam K 1985. Free jet scour below dams and flip buckets. *J. of Hyd. division ASCE, 111(2).*

Rae P. & Castro D.H. 1994. Modelling effects in the estimation of spillway free jet trajectory lengths. *Proc. ASCE national conference on hydraulic engineering, Buffalo.*

Torales M. et al. 1994. Itaipu spillway deterioration and maintenance after 10 years of operation. *Proc. ICOLD Q.71, R.38, Durban.*

U.S. Bureau of reclamation.1977. *Design of small dams.* Washington: Water resources publication.

Vischer D.L. & W.H.Hager.1998. *Dam Hydraulics.* Chichester: John Wiley & Sons.

Water Research Center.2002.Hydraulic studies of Agh Chay spillway & plunge pool. Un published Report. (In Farsi).

Water Research Center.1995.Hydraulic studies of the general model of Karun 3 spillway. Report No. 164. (In Farsi).

Water Research Center.1993.Hydraulic model studies for plunge pool banks protection of Karun 1 spillway with scale 1:62.5 (In Farsi).

Water Research Center.1986.Hydraulic model studies for spillway & plunge pool of Marun dam. Report No. 86. (In Farsi).

Whittaker, J.G. & Schleiss,A.1984. Scour related to energy dissipaters for high dam structures. Mitt nr 73.VAW/ETH, Zurich.

Yildiz, D. & Uzucek, E. 1994a. Experience gained in Turkey on scours occurred downstream of spillways of high head dams and protective measures. *Proc. ICOLD Q.71, R.9, Durban.*

Yildiz, D. & Uzucek, E.1994b. Prediction of scour depth from free falling flip bucket jets. *Int. water power & dam construction, Nov.*

Physical modeling and scale effects

(Courtesy of E. Bollaert, EPFL)

Rock Scour due to falling High-velocity Jets - Schleiss & Bollaert (eds)
© 2002 Swets & Zeitlinger, Lisse, ISBN 90 5809 518 5

Drag of emergent and submerged rectangular obstacles in turbulent flow above bedrock surface

P.A. Carling,
Department of Geography, University of Southampton, Highfield, Southampton, UK

M. Hoffmann, A.S. Blatter, A. Dittrich
Institut für Wasserwirtschaft und Kulturtechnik, Universität Fridericiana zu Karlsruhe, Karlsruhe, Deutschland

ABSTRACT: Flume experiments quantify the force balance equation for initial motion of isolated blocks. Experiments: (1) Initial motion on smooth concrete surface; (2) Initial motion on simulated rough bedrock. Three cases of initial motion: (1) the block is emergent and pivots; (2) the block is emergent and slides; (3) the block is submerged and slides. The drag coefficient (C_d) on rough bedrock was related to relative water depth and particle orientation. For submerged cases $C_d = 0.95$ for sliding, but for near emergent/emergent conditions C_d increases rapidly as relative depth reduces. For shallow water and emergent conditions large values of C_d pertain as the relative depth decreases and blocks pivot. For emergence, the lift coefficient (C_l) is zero. For submerged conditions negative C_d decreased primarily as relative depth increased (with block orientation being a minor additional factor) and assumed a constant positive value of about 0.07 on rough bedrock and 0.17 on smooth bedrock for relative depths $\cong 4$. Thus $C_l \cong 0.12 C_d$. Due to the turning moment, the friction coefficient (μ_f) for the smooth bed was not constant, but on the bedrock μ_f was constant, only varying between 0.5 and 0.8 depending on the lithology/degree of contact between bed/block.

1 INTRODUCTION

Isolated or grouped large boulders are commonly observed in bedrock channels. Some of these boulders may be residual 'lag' deposits that have fallen from the side walls of the channels and are too large to have been transported by floods. Other boulders may clearly have been transported and deposited by floods. Thus, in principle, knowledge of the forces required to move these latter boulders should allow estimation of the competent discharge, whilst the lag boulders should provide an upper limit to estimations of competence.

The question arises as to why competence studies particular to bedrock channels are required, when the entrainment of well-rounded cobbles from gravel beds consisting of mixed grain-size has been well-studied. The latter condition usually entails initial motion by a pivoting mechanism in relatively deep sub-critical flow (Komar 1996) because gravel-bedded rivers are free to adjust their boundaries such that negative feedback ensures super-critical and critical flows occur only locally (Grant 1997). In contrast within bedrock channels the bed slope and geometry of many channel reaches may be adapted to maximize the unit discharge through a minimization of the mean specific energy (Grant 1997, Tinkler 1997, Chanson 1999). This condition occurs when the flow is critical (Froude number ~ 1.0). However, the boundaries of many bedrock channels adjust slowly, if at all, to the flow regime and so, unlike gravel-bedded streams, the largest residual boulders in a bedrock channel may be subject to sub-critical, critical or super-critical flows. Consequently the largest water-transported boulders may have been moved by near-critical or super-critical flows. Finally, the bed of bedrock channels may be physically-smooth and boulders may be tabular rather than rounded. Thus a consideration of the mechanics of entrainment by sliding (as well as pivoting) in critical and super-critical flow is important in bedrock systems. In particular hydraulic jumps may be important, generating vibration of blocks and pressure between the base of blocks and the bed, or increased pressure on the top of blocks.

Boulders often lie on extensive smooth or rugose bedrock surfaces. Boulders may be either well-rounded and entrained by pivoting or they are blocky with beveled edges. In the latter case, initial motion may be by pivoting or by sliding. Boulders may be completely submerged or boulders may have dimensions comparable with, or greater than, the water depth. Thus the depth of flow above the obstacle may be small. The flow may exhibit free-surface undulations, and/or hydraulic jumps, which generate quasi-periodic temporal variation in the bed shear stress (Montes & Chanson 1998, Chanson 1999), or the boulder may be emergent such that large standing waves occur around and above the obstruction. For the emergent case, air is entrained both on the upstream side, along the lateral margins and on the downstream side of the obstacle. In the extreme, on the downstream side, the rockbed surface is exposed within a horseshoe-shaped standing wave or hydraulic jump (see Carling et al. 2002). Thus, in near-emergent or emergent cases, the effects of sharp-edges of boulders, stagnation zones, wake, air entrainment and wave drag all may influence the over-all drag on the obstacle.

The literature describing the parameters within the force balance equation for initial motion shows that the drag coefficient (C_d) and lift coefficient (C_l) of obstacles, and the friction coefficient (μ_f) between blocks and bedrock are poorly quantified for immersed or emergent boulders. The presence of an obstacle within the flow introduces local flow distortions that complicate an exact solution for the drag forces. This is especially true for shallow flows and emergent obstacles. Consequently in this study there are two objectives. Firstly, to provide guidance on the typical (if not exact) values of drag coefficients for regular blocks and secondly to determine the parameter combinations for which the flow is essentially uniform. For the latter condition the values of drag coefficients derived from this study then may usefully be compared with other detailed investigations of drag induced by rectangular blocks. The criteria for determining the relative degree of submergence for which flow in a wide channel become essentially uniform is that the lift coefficient should be small whilst the drag coefficient becomes independent of the Froude number, relative submergence, and relative width. Given these conditions, dynamic similarity pertains for regular shapes for a given Reynolds number (Flammer et al. 1970). In the present experiments the effects of the Froude number and relative width were eliminated, so that the effects of relative submergence on lift and drag could be examined.

2 DIMENSIONAL ANALYSIS

The important variables affecting the drag on single smooth-faced cubes or oblongs in flow with free surface effects are:

$$F = \phi(\overline{U}, h, H, L, B, \mu, \rho, g, W), \tag{1}$$

in which F is the drag force, $\overline{U}$ is the mean flow velocity, h is the water depth, H is the obstacle height, L is the length of the obstacle parallel to the flow, B is the breadth of the obstacle normal to the flow, μ the fluid viscosity, ρ is the fluid density, g is the acceleration due to gravity and W is the channel width (Fig. 1).

By dimensional analysis these variables can be reduced to the following dimensionless parameters:

$$C_d = \phi(R_e, Fr, h/H, h/L, W/B,), \tag{2}$$

where C_d is the drag coefficient, $R_e = \overline{U} h/v$ is the Reynolds number, $Fr = \overline{U}/\sqrt{(gh)}$ is the Froude number and $v = \mu/\rho$ is the kinematic viscosity. The velocity profile parameter P accounts for non-logarithmic profile effects. Other than selecting an appropriate reference level

84

for $\overline{U}$ in the flow depth, the effect of non-logarithmic velocity profiles usually can be neglected (Yen 1965, Flammer et al. 1970, Roberson & Chen, 1970).

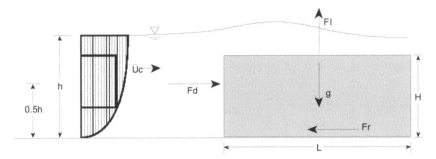

Figure 1. Definition of key variables.

3 EXPERIMENTAL APPARATUS AND METHOD

Two bed conditions were investigated within a tilting laboratory hydraulic flume. The flume has a working section of 10.56m and a width of 0.9m. Bed condition (i) consisted of lining the complete length of the flume with smooth-faced concrete paving stones 0.89m wide by 0.5m long. The paving stones were carefully grouted in place to provide a 'smooth' bed surface. Bed condition (ii) used simulated natural-stone paving stones with irregular surfaces that included 1mm to 3mm high flow-transverse rugosities and ridges. Cubes and oblongs of varying dimensions and weights were machine-sawn from sandstone or slate (Table 1).

Despite flow velocities typically between one and two meters per second, the physically smooth beds resulted in the Reynolds number for the majority of flows being low and typically just within the rough-turbulent regime (Re >400 to 500: Chang 1988, Dingman 1984). For these conditions the usual relationship for the logarithmic layer in the rough turbulent regime (*e.g.* Nikora & Smart 1997) was fit to the velocity data:

$$U_z = \frac{u_*}{k} \ln \frac{z}{z_o} \ , \qquad (3)$$

where z is the height above the bed and k is von Kármán's constant (in clear water = 0.40) and $u_* = \sqrt{ghS}$ where S is the water surface slope in uniform flow. A few test runs considered transitional conditions and, in the latter case, the depth-mean velocity was derived using a log-normal function for hydraulically transitional flow (Zanke 1982).

During experiments for bed condition (i) a miniature impeller flow meter was used to obtain representative log-normal velocity profiles by measuring the flow speed at millimeter to centimeter increments in the vertical for 14 locations in the vertical. The goodness of fit of Equations 3 and 4 to the data, as measured by the r^2 value of a least-squares regression analysis was in excess of 0.99. For individual tests the velocity was also measured at half the obstacle height and at 0.6 of the depth. During experiments for bed condition (ii) an acoustic doppler velocimeter (ADV) was used. Both instruments were calibrated and there was no statistical difference when comparing the average velocity values obtained using either instrument. Manometers recorded water surface slope and water depth was measured using a point gauge and millimeter-graduated scales on the glass-side walls of the flume. Video and still photography were used to record the style of initial motion for blocks.

Owing to the basic nature of the experiments and the complicated nature of initial motion, the analysis is restricted to a simple evaluation of the force balance (Graf, 1971). The basic relations are given as:

$$F_d = K_1 C_d dB \frac{\rho \overline{U}_c^2}{2} \qquad \text{(drag force)} \qquad (4)$$

$$F_l = K_2 C_l BL \frac{\rho \overline{U}_c^2}{2} \qquad \text{(lift force)} \qquad (5)$$

$$W = K_3 (\rho_s - \rho) gHBL \qquad \text{(submerged weight)} \qquad (6)$$

$$\mu_f = \frac{W \sin \alpha + F_d}{W \cos \alpha - F_l} \qquad \text{(friction coefficient)} \qquad (7)$$

wherein α is the bed angle and K_1, K_2 and K_3 are parameters which subsume shape effects for non-perfect regular blocks and variation in the definition of the critical velocity. The latter point is considered later.

Substituting 4 through 6 into 7 yields:

$$\frac{\overline{U}_c^2}{(\rho_s / \rho - 1) g \sqrt{HL}} = \frac{2 K_3 (\mu_f \cos \alpha - \sin \alpha)}{K_1 C_d \sqrt{H/L} + \mu_f K_2 C_l \sqrt{L/H}} , \qquad (8)$$

Within Equation 8 it should be noted that the breadth of the block transverse to the flow is not an explicit factor in the force balance. This point is referred to below.

If the flow-normal area (A_d) and the volume of the block (K) are introduced as independent variables, then Equations 5 through 7 can be rewritten as:

$$F_d = K_1 C_d A_d \frac{\rho \overline{U}_c^2}{2} , \qquad (9)$$

$$F_l = K_2 C_l L \frac{A_d}{d} \frac{\rho \overline{U}_c^2}{2} , \qquad (10)$$

and,

$$W = (\rho_s - \rho) gK . \qquad (11)$$

Assuming $\cos \alpha \approx 1$, $\sin \alpha \approx 0$, $K_1 = 1$ and $K_2 = 1$, then by substitution the drag coefficient is:

$$C_d = \mu_f \frac{2g(K\rho_s - A_b h\rho)}{(\rho A_d) \overline{U}_c^2} , \qquad (12)$$

and the lift coefficient is:

$$C_l = \frac{2gKh(\rho_s - \rho)}{LA_d\overline{U}_c^{\ 2}} - \frac{hC_d}{L\mu_f}. \tag{13}$$

wherein ρ_s and ρ are the densities of the blocks and the water respectively. K is the volume of the block, A_b and A_d are the basal and the flow-normal submerged areas of the blocks respectively. The velocity U_c appropriately should be the current speed measured at half the block height for incipient motion, but latterly values of drag and lift are reported calculated using the depth averaged velocities for reasons explained below. It should be noted, firstly, that Equation 12 assumes the blocks move only by sliding and the lift force is neglected and, secondly, that the determination of C_l using Equation 13 utilizes C_d as obtained from Equation 12. Thus these equations apply only to static blocks immediately prior to motion by sliding. Some blocks were entrained by pivoting, as is explained later, but no determinations of the drag coefficients for entrainment by pivoting were conducted; this is because the coefficients would be variable with time as the attitude of the blocks would vary during the pivoting process.

For the flume tests the friction coefficient μ_f was calculated for each rock type by first assuming C_d was known, using an engineering standard for regular blocks (settling through a water column) with varying combinations of the lengths: D_a D_b, D_c (Rosemeier, 1976). The average value of μ_f for each rock type was then used in Equation 12 to derive specific values of C_d for each test condition that included submerged or emergent conditions. Latterly these calculated values of the friction coefficient were found to compare well with values derived directly from laboratory tests. In the latter case, values of μ_f were obtained by pushing or pulling blocks across the test beds using a digital force gauge. The same procedure was used on a relatively smooth limestone bedrock surface in a natural river channel.

As noted above, the value of drag coefficient depends on the characteristic velocity used to calculate C_d (Flammer et al., 1970). The velocity (U) integrated over the height of the obstacle will give a larger C_d value than one using $\overline{U}$ measured above, or integrated over, the height of the boundary layer. We used both approaches, but for our shallow flows, there was no statistically significant difference in the results. Thus we report data pertaining to the depth-average velocity, which is the parameter of interest for discharge estimates. The above procedure was justified in-as-much as the method obtained values of the drag coefficient for submerged blocks, which are consistent with published literature on the value of the drag coefficient for regular blocks adjacent to a boundary. Thus it can be reasonably assumed that the C_d values for emergent blocks are consistent and representative.

Finally the relative turning moments for pivoting blocks can be defined as:

$$\frac{M_1}{M_2} = \frac{F_N\frac{L}{2}}{F_D\frac{H}{2}} = \frac{W\cos\alpha L}{W\sin\alpha H} = \frac{1}{\tan\alpha}\frac{L}{H}, \tag{14}$$

where F_N is the normal force, F_D is the drag force, W is effective weight of the block , M_1 is the holding moment, M_2 is the driving moment and α is the slope of the channel. As the ratio M_1/M_2 increases there is a greater propensity for the blocks to slide from rest rather than pivot.

4 EXPERIMENTAL PROCEDURE
The procedure consisted of placing a given block on the flume bed with either a smooth cut-face normal to the approach flow, or at a 45° angle such that in the latter case the flow impinged on a vertical block corner. In each case for repeat tests, the blocks were placed in exactly the same

position with reference to a location mark on the bed. For bed condition (i) there was one location. However bed roughness varied slightly for bed condition (ii) and so tests were replicated at two different locations. Blocks of various size (Table 1) were placed singly in the center of the flume such that W/B ranged between 3.38 and 90. For sub-critical flow Flammer et al. (1970) have shown that W/B must exceed 20 to prevent flow blocking between obstacle and side-wall being a factor affecting the backwater curve and hence the drag coefficient. For super-critical flow they showed that W/B reduced to 2.0 because the obstacle- induced flow disturbance cannot propagate upstream and the disturbance waves extend downstream as increasingly narrow wake angles. Thus our experiments largely eliminated the factor W/B.

Flow was adjusted using the discharge control, tailgate and bed slope such that the approach flow was uniform with a Froude number in the region of 1.5 to 2.5. By careful adjustment, flow conditions were obtained such that the drag coefficient C_d was independent of the Froude number. With this experimental design the variation in the drag can be described primarily as a function of relative submergence (h/H) and, to a lesser extent, block orientation (L/B).

By adjusting flume slope and tailgate height, flow depths were less than, equal to, or greater than the block height. Discharge was increased slowly until the blocks moved slightly. Video-frames obtained immediately prior to initial motion provided a visual qualitative representation of the style of flow around the obstacles. A qualitative description of the style of initial motion was recorded and then the block was removed from the flume. With the block removed, the velocity, depth, water surface and bed slopes were recorded. Thus the incident flow conditions associated with initial motion were recorded but the local effects of the presence of the obstacle were not accounted.

5 RESULTS AND DISCUSSION

Values of z_o the roughness length were typically 0.88mm to 1.1mm. Given $k_s \cong 30z_o = 2$ to 3mm then the hydraulic bed roughness was similar to the physical height of the small rugosities and transverse steps in the simulated bedrock. The drag coefficient, as anticipated, was independent of the Froude number and Reynolds number.

Variation in wave-type.- Three styles of water surface instabilities associated with emergent blocks were recorded in accord with examples of waveforms associated with shallow flow past obstacles on a plain non-mobile bed (see detail in Zgheib & Urroz-Aguirre 1990).

Variation in style of initial motion.- Three different cases of initial motion were recorded: (1) The block is emergent and pivots out of place; (2) The block is emergent and slides out of place; (3) The block is submerged and slides out of place. Initial motion by pivoting was always induced by flow characteristics associated with emergence.

Variation in the drag coefficient.- For practical applications the drag coefficient for incipient motion was shown to be a function of relative submergence with a minor contribution from block orientation; both factors being readily estimated from field data (Figs 2 & 3). Although Equation 8 demonstrates that block width (B) is not an explicit factor in defining the drag coefficient a weak effect of flow blocking was recorded. A least squares empirical function takes this latter effect into account, and applies equally well to both emergent and fully submerged blocks on the simulated bedrock surfaces. For the smooth surface:

$$C_d = \left(\frac{0.44}{(h/H)(L/B)} \right) + 1.01 \qquad\qquad r^2 = 0.80 \qquad\qquad (15)$$

For the rough surface:

$$C_d = \left(\frac{0.51}{(h/H)(L/B)} \right) + 0.95 \qquad\qquad r^2 = 0.76 \qquad\qquad (16)$$

where B is the block length transverse to the flow and L is the length parallel to the flow. The uncertainty expressed by the rough surface data-spread is described by limiting curves; $y = (0.8/x) + 1.3$ and $y = (0.4/x) + 0.6$. For relative depths of about 2, or greater, C_d assumes a constant average value of about 0.95, for the rough test bed and 1.0 for the smooth test bed; which values compare well with other recommendations (Hoerner 1965, Roberson & Chen 1970, Carling & Grodek 1994, Wende 1999, Carling & Tinkler 1998). However for shallow submergence the drag coefficient can be greater than 1 and increases very rapidly as water level falls below the block height.

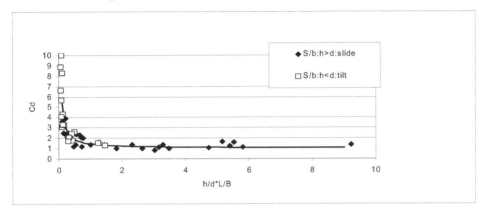

Figure 2. Variation of the drag coefficient, for blocks on smooth concrete bed (S/b), as a function of the relative depth and block form factor.

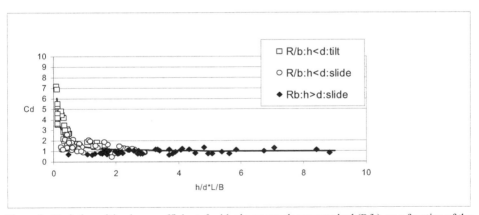

Figure 3. Variation of the drag coefficient, for blocks on rough concrete bed (R/b), as a function of the relative depth and block form factor.

Variation in the lift coefficient.- Estimates of C_l are scattered but show that for greater relative submergence the lift coefficient tends to a constant value; of the order of ten percent of the drag coefficient, and is greater for the smooth concrete surface when compared with the rough surface (Fig. 4). Lift can only occur for submerged blocks and so h/H cannot be less than unity. The existence of two data points for $(h/H)(L/B)$ less than one demonstrates that for shallow flows, flow blocking ($B>L$) may induce a standing wave above the block and induce weak negative lift. Negative lift is only associated with shallow flow and associated flow blocking.

As relative depth increases, and blocking reduces, lift values tend to become positive. The data trends may be described by least-squares regression analyses constrained to a constant value of lift for critical values of $(h/H)(L/B)$ greater than a given value:

$$C_l = 1.0098[2 - (h/H)(L/B)] - 0.2633[2 - (h/H)(L/B)]^2 - 1.1762 \qquad \text{Smooth bed} \quad (17)$$

$$C_l = 0.0198[4 - (h/H)(L/B)] - 0.0347[4 - (h/H)(L/B)]^2 + 0.073 \qquad \text{Rough bed} \quad (18)$$

Equations 17 and 18 are specific to the critical $(h/H)(L/B)$ values of 2 for the smooth bed and 4 for the rough bed. At these points the lift over the smooth and rough beds is deemed to become constant, equal to the average values of lift ($C_l = 0.1717$ and 0.0734 respectively) for the data above the critical value. The critical values were chosen arbitrarily following inspection of the data plot, but are consistent with other studies. For example, Flammer et al. (1970) selected critical values of (h/H) of 1.6 and 4 respectively to delimit the region of moderate free surface effects from the regions of pronounced and negligible effects respectively. Thus for practical applications the value of L/B in Equations 15 through 18 may be set to unity.

Variation in the coefficient of friction.- Figure 5 shows that the coefficient of friction for blocks on the rough simulated bedrock surface is approximately constant and independent of the turning moment when the entrainment force is applied normal to the flow-transverse face in an axial position. Each block was orientated with the long and intermediate axes parallel and transverse to flow respectively. The increase in the width of block 5 compared with block 6 ensured that the edges of the block were less likely to interact with small bed rugosities thus reducing the overall coefficient of friction. However lengthen a block (Block 7) caused the friction coefficient to increase. In contrast, on the smooth simulated bedrock surface the coefficient of friction is greater than that shown in Fig. 5. This is because the absence of rugosities ensures greater contact between the block and bedrock surfaces. In Fig. 6 the force was applied normal to the upstream block face but by varying the height of the applied force the value of the turning moment could be varied systematically. This kind of experiment reproduces the variation in applied force such as that occurring naturally when turbulent flow pulses, moving towards or away from the bed, make contact with the face of the obstacle at acute angles. As the turning moment increases, a blocks begins to pivot about the downstream edge such that the base surface is no longer in contact with the bed and the friction coefficient decreases. The friction coefficient is larger for those blocks of small basal area compared with the block height (Fig. 6). Blocks placed on the bed with the long axis vertical are prone to entrainment by pivoting, whilst blocks placed with the long axis transverse to or parallel with the flow are entrained by sliding.

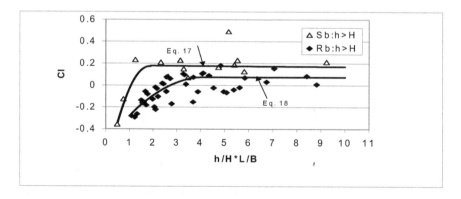

Figure 4. Variation of the lift coefficient, for blocks on smooth (S/b) and rough concrete bed (R/b), as a function of the relative depth and block form factor.

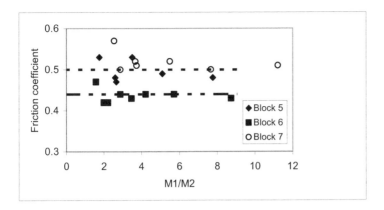

Figure 5. Coefficient of friction, on rough concrete bed, as a function of the turing moment.

Points represent averages of 150 to 250 measurements for each block. The force was applied at the center downstream face by pulling the block in a direction normal to the face using a high-tensile wire attached to a force gauge. Curves shown represent the average values for blocks 5 and 6.

In the field a variety of sawn and natural blocks pulled across smooth limestone bedrock showed evidence for a constant value of the friction coefficient, or a slight reduction as the turning moment increased (Fig. 7).

In summary, for the smooth concrete bed, with maximum contact between block and bed, μ_f was 0.80 for sandstone and 0.60 for slate. For the simulated bedrock the number of points of contact were reduced such that μ_f equalled 0.50 and 0.59 for the two rock types respectively. In a natural smooth-bedded limestone channel μ_f was 0.81 for sandstone blocks, 0.73 for slate and 0.70 for natural limestone blocks derived from the channel bed.

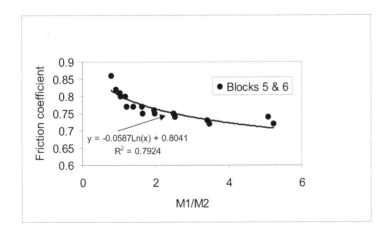

Figure 6. Coefficient of friction, on smooth concrete surface, as a function of the turning moment. Points represent averages of 150 to 250 measurements for each block. The force was applied at the centre downstream face by pulling the block in a direction normal to the face using a high-tensile rod attached to a force gauge.

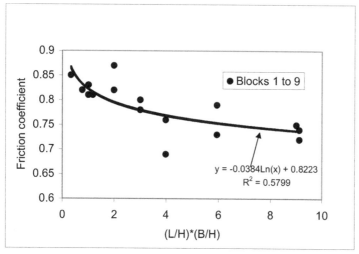

Figure 7. Variation of the coefficient of friction, on smooth concrete surface, as the basal area of the block increases relative to the obstacle height.

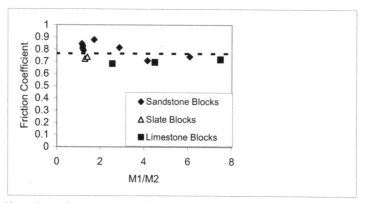

Figure 8. Variation of the coefficient of friction, as a function of the turning moment, on smooth natural limestone forming the bed of the River Dee, northern England. Curve shown represents the average value of all the data.

6 CONCLUSIONS

Three different cases of initial motion were recorded: (1) the block is emergent and pivots out of place; (2) the block is emergent and slides out of place; (3) the block is submerged and slides out of place. Thus initial motion of regular blocks by pivoting on a smooth bedrock surface is induced by the flow characteristics associated with emergence.

The variation in the drag coefficient (C_d) for conditions of initial motion on rough bedrock was strongly related to relative water depth (h/H) and weakly related to particle orientation (L/B). For the fully submerged case $C_d = 0.95$, but for near emergent and emergent conditions C_d increased rapidly as relative depth was reduced. Setting C_d equal to 0.95 is appropriate where blocks were deeply submerged ($h/H > 2$) and entrained by sliding. However where blocks were in shallow water or emergent ($h/H < 2$) increasingly large values of C_d pertain as h/H decreased. Often, in the latter case, blocks were entrained by pivoting.

For conditions of emergence, the variation in the lift coefficient (C_l) was restricted to negative values associated with flow blocking by oblongs placed with the long axis transverse to the flow. For submerged conditions ($h/H \cong 2$ to 4) negative lift coefficients decreased in value as $(h/H)(L/B)$ increased and assumed a constant positive value of about 0.07 on rough bedrock and 0.17 on smooth bedrock. These latter values indicate that $C_l \cong 0.12 C_d$.

Due to the turning moment, the friction coefficient (μ_f) for the smooth bed was not constant, but on the simulated bedrock surface the friction coefficient was approximately constant. Values of μ_f varied between 0.5 and 0.8 depending on the lithology and the consequent degree of contact between bed and block surfaces. It may be concluded that the friction coefficient is weakly and positively dependent on the height of the block relative to the basal area but for blocks of small height relative to the basal area μ_f is effectively constant.

7 ACKNOWLEDGEMENTS

The authors would like to thank Mssrs. Andy Quin, Paul Williams and Charles Blakeley (Lancaster University) for assistance in preparing the flume. The constructive comments of Drs. Mark Schmeeckle (Florida State University) and Zhixian Cao (Southampton University) improved the presentation of the manuscript.

REFERENCES

Carling, P.A. and Grodek, T. 1994. Indirect estimation of ungauged peak discharges in a bedrock channel with reference to design discharge selection. *Hydrological Processes* 8: 497-511.

Carling, P.A., Hoffmann, M. & Blatter, A.S. (2002) Initial motion of boulders in bedrock channels. pp 147-160 In: House, K. (ed.) *Ancient Floods, Modern Hazards: Principles and Applications of Paleoflood Hydrology*, Water Science and Applications Vol. 5, American Geophysical Union.

Carling, P.A. and Tinkler, K. 1999. Conditions for the entrainment of cuboid boulders in bedrock streams: an historical review of literature with respect to recent investigations, pp 19-34 In: K.J. Tinkler and E. E. Wohl (eds.) *Rivers Over Rock*, Geophysical Monograph 107, Washington, USA: American Geophysical Union.

Chang, H.H. 1988. *Fluvial Processes in Engineering*, Chichester: Wiley, 432pp.

Chanson, H. 1999. Comment on "Critical flow constrains flow hydraulics in mobile-bed streams: A new hypothesis" by G. E. Grant. *Water Resources Research* 35: 903-905.

Dingman, S.L. 1984. *Fluvial Hydrology*, New York: W.H. Freeman, 383.

Graf, W.H. 1971. *Hydraulics of Sediment Transport*, New York: McGraw-Hill 513pp.

Grant, G.E. 1997. Critical flow constrains flow hydraulics in mobile-bed streams. *Water Resources Research* 33: 349-358.

Flammer, G.H., Tullis, J.P. and Mason, E.S. 1970. Free surface, velocity gradient flow past hemispheres. *J. Hydraulics Division, ASCE,* 96: 1485-1502.

Hoerner, S.F. 1965. *Fluid Dynamic Drag*. Bricktown, NJ: Hoerner Fluid Dynamics, 3-17.

James, C.S. 1990. Prediction of entrainment conditions for nonuniform, noncohesive sediments. *J. Hydraulic Research* 28: 25-41.

Komar, P.D. 1996. Entrainment of sediments from deposits of mixed grain sizes and densities. In P.A. Carling and M.R. Dawson (eds.), *Advances in Fluvial Dynamics and Stratigraphy* pp 127-181, Chichester: Wiley.

Montes, J.S. and Chanson, H. 1998. Characteristics of undular hydraulic jumps: experiments and analysis. *J. of Hydraulic Engineering* 124: 192-204.

Naudascher, E. 1992. Hydraulik der Gerinne und Gerinnebauwerke, 2, Heildelberg: Springer-Verlag.

Nikora, V.I. and Smart, G.M. 1997. Turbulence characteristics of New Zealand gravel-bed rivers. *J. Hydraulic Engineering, ASCE* 123 (9): 764-773.

Roberson, J.A. and Chen, C.K. 1970. Flow in conduits with low roughness concentration. *J. Hydraulics Division, ASCE,* 96: 941-957.

Rosemeier, G. 1976. *Winddruckprobleme bei Bauwerken*, Heidelberg: Springer-Verlag.

Tinkler, K. 1997. Critical flow in rockbed streams with estimated values for Manning's *n*. *Geomorphology* 20: 147-164.

Wende, R. 1999. Boulder bedforms in jointed-bedrock channels, In A.J. Miller and A. Gupta (eds), *Varieties of Fluvial Form*, 189-216, Chichester: Wiley.

Yen, B-C. 1965. Discussion of: Large-scale roughness in open channel flow. *J. Hydraulics Division, ASCE*, 91, 257-262.

Zanke, U. 1982. Grundlageder Sedimentbewegung. Heidelberg: Springer-Verlag.

Zgheib, P. W. and Urroz-Aguirre, G.E. 1990. Flow transition around a single large bed element. In H.H. Chang and J.C. Hill (eds) *Hydraulic Engineering* 2: 1036-1041, Proceeding of the 1990 National Conference of the ASCE.

Block Number	Dry Weight M (g)	Wet Weight M (g)	Long Axis D_a (mm)	Int. Axis D_b (mm)	Short Axis D_c (mm)	Rock Type
1	288	295	50	50	50	Sandstone
2	583	596	100	50	50	Sandstone
3	874	894	150	50	50	Sandstone
4	1169	1196	199	50	50	Sandstone
5	1745	1786	150	99	50	Sandstone
6	2637	2696	150	150	50	Sandstone
7	2668	2689	230	99	50	Sandstone
8	4380	4420	123	123	123	Sandstone
9	19660	19720	266	177	177	Sandstone
10	397	397	51	51	51	Slate
11	5570	5570	124	124	124	Slate
12	16820	16820	178	178	178	Slate

Table 1: Summary of experimental materials. Machine-sawn slightly porous blocks 1 through 12 were weighed air-dried and after immersion in water for several days.

Parametric analysis of the ultimate scour and mean dynamic pressures at plunge pools

L.G. Castillo E.

Technical University of Cartagena, Spain

ABSTRACT: Studies to find out the actions at dam toes of impinging jets have been carried out using two different ways of research: the study of scour and the study of pressures. The instrumentation and measurement objectives have marked the difference between the two ways of research and "apparently" the type of formulation obtained. In this paper the general formulation of ultimate scour and ultimate pressures is presented. Their correspondence is outlined and reveals a unique type of formulation. A practical parameter, called the "incremental energy dissipation", is presented in order to estimate the resistance of the plunge pool bottom that is necessary to withstand the power of the jet. This parameter also allows an analysis of the corresponding pressure fluctuations.

1 INTRODUCTION

The jet action at the dam toe was carried out following two different ways: the study of plunge pool scour and the study of instantaneous, mean and fluctuating pressures. The instrumentation and the objective of the measurements marked the difference between these two approaches and, as a consequence, "apparently" the type of formulation obtained. Thus, in the scour formulations, the main measurement is the depth and shape of the scour hole, while in the second approaches the main objective is the characterization of the pressure on the plunge pool floor.

The classical scour formulae constitute empirical correlations obtained by model tests, such as Schoklitsch (1932), Veronese (1937), Muñoz (1964), Rajaratnam (1981), Rajaratnam et al. (1995), or obtained by prototype observations, such as Wu (1973). Combined prototype-model expressions are the one by Mason et al. (1985) and the one by Lopardo et al. (1987). Ramírez et al. (1990) carried out a summarizing study of the different scour formulae by means of turbulent jet theory.

As far as the pressure characterization studies are concerned, the main empirical formulations have been determined exclusively in models, because of the complexity of installing instrumentation in prototypes. These formulations have evolved since the work of Moore (1943), Lencastre (1961), Cola (1965), Aki (1969), Hartung & Häusler (1973), Beltaos (1976) until the more recent work of Xu-Do-Ming et al. (1983), Lemos et al. (1984), Cui Guang Tao et al. (1985), Ervine & Falvey (1987), Withers (1991), Ervine et al. (1997) and the research program of the Hydraulics Laboratory of the Universidad Politécnica de Cataluña (UPC): Castillo (1989), (1990), Castillo et al. (1991), Armengou (1991), Puertas (1994), Castillo et al. (1996) and Castillo et al. (1999).

In this paper some of the general formulations of ultimate scour and pressures are presented, a correspondence is outlined and a single type of formulation is defined. In addition, a practical parameter, called the "incremental energy dissipation", is presented in order to estimate the resistance of the plunge pool bottom that is necessary to withstand the power of the water jet. This allows the analysis of the different actions in the plunge pool.

2 ANALYSIS OF JET IMPINGEMENT

At overflow spillways, the total energy is dissipated by the following mechanism (see Fig. 1): firstly by friction, spreading and air entrainment for jet velocities higher than 6 m/s (fall heights of more than 1.83 m); and by the atomization of the water for jet velocities higher than 20-30 m/s (heights of 20-46 m). Secondly, if there is a rigid plunge pool bottom (Fig. 1a), the combined effect of diffusion and impact and the internal friction of a submerged hydraulic jump dissipate the jet energy.

For a scour hole (Fig. 1b), the ultimate state of scour is time-independent due to the total jet diffusion. Scour hole formation is the main mechanism, until the velocity of the jet diminishes to the level where the shear stress on the bed reaches a critical value according to Shields (not necessarily numerically equal) [Ramírez et al. (1990)].

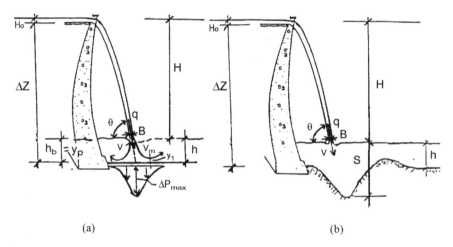

(a)   (b)

Figure 1. Plunge pool at dam toe: (a) Concrete lined floor (rigid plate), (b) Scour hole.

That means that the scour will grow on until a water cushion is obtained which is sufficient for the turbulent jet to be diffused totally. In addition, the velocities and pressure fluctuations have diminished to a certain level, in such a way that they are unable to mobilize and extract materials from the bed. For this fully developed jet the following ratio could be applied:

$$\frac{S}{B \sin \theta} = S^* = K \left(\frac{V}{V^*_c} \right)^n = V^{+n} \tag{1}$$

where S = scour depth from the water level to the deepest point of the plunge pool; B = thickness of the impacting jet; θ = angle of the jet impingement; V = the velocity of the impacting jet; V^*_c = the critical friction velocity (shear stress threshold) so that the movement of particles is produced in the plunge pool; K and n are coefficients which depend on the shape of the jet and which have a theoretical value of K = 2.7 and n = 2 for a plane jet and K = 6.4 and n = 1 for a circular jet [Tennekes & Lumley (1972)].

2.1 Theoretical discussion and joining of the ultimate scour formulae

Most scour formulae obey the following common form:

$$S = K \frac{H^\alpha q^\beta h^\delta}{D_i^\gamma} \tag{2}$$

where H = total height up to downstream level; h = depth of initial water cushion; D_i = representative diameter of the particles which remain in the plunge pool and K, α, β, γ and δ are coefficients which the different authors have obtained experimentally.

Ramírez et al. (1990), based on the vertical free-fall configuration and in conjunction with the submerged turbulent jet theory, obtained a ratio which would be valid when the particle size no longer plays a part in the phenomenon:

$$S^* = \frac{S}{B} = 2.34 L^{+3/2} \tag{3}$$

where $L^+ = V/(qg)^{1/3} = V/V_c$; q = specific flow; g = gravity acceleration; V_c = critical velocity.

Thus, by means of a suitable reduction to common parameters, they carry out a comparison of the empirical scour formulae, classifying them in three general types: Type I (non dimensionally homogenous) and Type II (dimensionally homogenous), the particle size is concerned; Type III (dimensionally homogenous), the particle size is not concerned and the limit ratios are constituted according to Veronese (1937). Among the Type III limit formulations, Wu´s (1973) empirical non-dimensional ratios are K = 2.11; α = 0.235; β = 0.51; γ = 0 and δ = 0, obtained from prototypes and those of Lopardo et al. (1987) are K = 2.50; α = 0.25; β = 0.50; γ = 0 and δ = 0, obtained from models and prototypes. Table 1 shows the coefficients corresponding to theoretical formulations of plane and circular jets, as well as some limit scour formulations, expressed as a function of the variables S^* and L^+:

Table 1. Coefficients of general formulation $S^* = K L^{+n}$.

Author	K	n	Observation
Theoretical plane jet	2.70	2.00	Unlimited depth
Theoretical circular jet	6.40	1.00	Unlimited depth
Ramírez et al.	2.74	1.50	Plane. Maximum scour depth
Wu	1.79	1.47	Plane. Maximum scour depth
Lopardo et al.	2.10	1.50	Plane. Maximum scour depth

2.2 Mean dynamic pressure at the bottom of the plunge pool

The characterization of pressures due to jet impingement has been studied using different instruments (piezometers, pressure transducers, etc.), different media (air/air, air/water, water/water) and different jet geometries (rectangular nappe jet or circular jet). As far as the rectangular nappe is concerned, the results obtained by different authors may be summarized by the following general formulation, which represents the maximum dynamic pressure and its spatial distribution:

$$\Delta p_{max} = C\rho \left(\frac{V^2}{2}\right)\left(\frac{B}{h}\right) = \frac{Cq\gamma H^2}{h\left(2gH^3\right)^{1/2}} \tag{4}$$

$$\Delta p = \Delta p_{max} e^{-k|x/h|^2} \tag{5}$$

where ρ and γ are the density and the specific weight of the water. In Table 2, the coefficients C and k are shown, and a considerable dispersion of data can be seen, which is not surprising in view of the different natures of the tests.

The jet diameter D, or the jet thickness B, mainly depends on the type of discharge device (orifice jet or overflow nappe) and on the falling jet characteristics. Ervine et al.´s (1997) formula include an estimation of the jet spread due to the initial turbulent velocity in the case of an orifice jet. This formula is valid for h/D > 4-5 and for H/H_b (jet plunge length / jet break-up length) < 0.5. The exponent k of the pressure distribution formula in the radial direction is 25 for non-effective water cushions (h/D < 4) and 30 for effective water cushions.

However, there is still no estimation of jet spread in the case of an overflow nappe, while it can be seen in any case that the models that are based on Froude's similarity overestimate the pressures. For this reason, the determination of the thickness B is only carried out by gravitational considerations.

Table 2. Coefficients C and k of the general formulae of mean dynamic pressure

Author	C	k	Trial characteristics	Means	Observation
Cola	7.18	40.51	$B = 12 - 24$ mm $h = 0.165–0.835$ m $V_0 = 1{,}3 - 4.8$ m/s	Water Submerged jet with -out aeration. Symmetrical bidimensional	
Hartung & Hausler	5	19.6	Theoretical Jet. disintegration depth $y_k = 5B$	Water. Unlimited jet depth	If jet is considered rough, then coef. Erv. C=3.56; k=9.92
Beltaos:	8	42	$B = 0.224$ cm $h/B = 45.5 - 68.2$	Air. Bidimensional Different angles of impingement	Adjustment verification eq. Schauer &Eutis
Cui Guang et. al:	5.2–6.35	12.56	Model without scale Prototype Q=80 m³/s H =165 - 187m $h = 32.5 - 54.9$ m	Water. Bidimensional non symmetrycal (model of arch dam)	Possible effects of scale of trial unknown
Armengou:	3.19	25	$H = 1.8 - 5.5$ m $Q < 50$ l/s $h < 1.2$ m $V = 6 - 10.4$ m/s *$H/H_b = 0.4 - 2.73$	Water. Non symmetrical Bidimensional Aerated jet	First values in starting of experimental facility trials
Puertas:	3.88	2	H =1.85 - 5.45 m $Q < 86$ l/s $h = 0.08 – 0.80$ m $V = 6 - 10.4$ m/s *$H/H_b = 0.4 - 2.73$	Water No symmetrical Bidimensional Aerated jet	The exponent of eq. (5) is $m = 0.5$
Ervine et. al:	38.4(1-Ci)(D/h)	25-30	$H = 0.51 - 2.63$ m $Q < 63$ l/s $h = 0.10 - 0.5$ m $V = 4 - 25$ m/s **$H/H_b < 0.5$	Water Circular jet Aerated jet	In D considers term of lateral spread by turbulence

*H_b = jet break-up length. Nappe flow: Horeni (1956): $H_b \approx 6 \, q^{0.32}$

** Circular jet: Ervine et. al (1997): $H_b / D_0 F_0^2 = 1.05 / C^{0.82}$; $C = 1.14 T_u F_0^2$ = $\beta i/(1+\beta i)$ = impingement initial air concentration; $Tu = v'/V$ = jet turbulent intensity ; $\beta i = Qa/Qw = K_1 \left[1 - V_{min} / V\right] \sqrt{H / D}$; $K_1 = 0.2$ (smooth turbulent jet); $K_1 = 0.4$ (very rough turbulent jet); $V_{min} \approx 1$ m/s = minimum velocity for air entrainment; Qa = air entrainment rate; Qw = water discharge rate; D_0 = initial jet diameter; F_0 = initial Froude number;

$D = Dc + 2\varepsilon$; $\varepsilon = (1.14 T_u U_0^2 / g) \left[\sqrt{2 H_b /(D_0 F_0^2) + 1} - 1 \right]$; D = impingement jet; Dc = jet core; ε = lateral spread.

Puertas´s (1994) formulation covers in a global manner jet break-up lengths situated between $0.4 < H/H_b < 2.7$. For this reason, it possibly underestimates the mean dynamic pressure coefficient $C_p = (\Delta p_{max})/V^2/2g$. Castillo (1998) carried out a new analysis with the data of Puertas and proposed different formulations of $C_p = f\ (h/B,\ H/H_b)$. On the other hand, the Puertas´ formulation is only valid for an effective water cushion:

$$h_e \geq 0.368q^{0.50}H^{0,25} \tag{6}$$

If there is no effective cushion, the bottom pressure is only reduced by the jet friction with the air and the friction effect at the pool bottom. The pressure distribution equation (5) is very different from all the other formulations (m = 0.5 and k = 2) and this could reflect the possibility that the flow was mainly unidirectional (downstream) after impingement.

2.3 Analysis of the pressures for ineffective water cushions

In the tests it has been observed that the sensor receives the impact in a non-uniform manner, because of the oscillation and break-up of the nappe. In this way, the results without an effective cushion are expected to be more varied. An important fact, already noticed by Lencastre (1961) and verified by Castillo (1989) and Castillo et al. (1991), is that the maximum pressure fluctuation corresponded to small cushions of water and not to direct impact (h = 0); possibly one of the reasons might be the effect that is discussed here. However, the reason that was highlighted by those authors should not be discarded: ".... by the small or null effect of small water cushions in the energy dissipation and by the advantage that a certain water cushion thickness offers for the development of turbulence". In order to know which proportion of distortion corresponds to a "measurement error" and which to the phenomenon of turbulence, new and more widely reaching tests should be carried out, so that a general law of pressures for ineffective cushions could be drawn up.

Given that these phenomena are not considered in any of the formulations noted here, in the analysis as an approximate method for the case of direct impact, the formulation proposed by Moore (1943) is used, where the mean pressure is considered as the difference between the height of the discharge and the energy loss due to the friction effect of the pool depth y_p, so that:

$$\left(\frac{y_p}{y_c}\right)^2 = \left(\frac{y_1}{y_c}\right)^2 + 2\left(\frac{y_c}{y_1}\right) - 3 \tag{7}$$

in which y_1 = contracted depth at the toe of the spill and y_c = critical depth.

The energy loss ΔE, deduced from the momentum equation, is:

$$\Delta E = \Delta z + (3/2)y_c - y_1 - \frac{V_m^2}{2g} \tag{8}$$

where $V_m = (V/2)(1+\cos\theta)$ = velocity corresponding to depth y_1; Δz = height from overflow crest to floor of plunge pool; $\cos\theta = 1.06/(\Delta z/\ y_c +3/2)^{0.5}$; θ = angle of the jet impingement.

3 RELATION BETWEEN ULTIMATE SCOUR AND MEAN DYNAMIC PRESSURE

If in the equation of Puertas (1994) suitable transformations and calculations are carried out, the following formulation is obtained as a function of the variable $L^+ = V/(qg)^{1/3}$:

$$\frac{\Delta p_{max}}{B} = 2.74L^{+3/2}F_b \tag{9}$$

$$F_b = \left(\frac{B^{0.5}H^{0.5}}{h} \right) = \frac{1}{(2g)^{0.25}} \frac{q^{0.5}H^{0.25}}{h} = 0.475 \frac{q^{0.5}H^{0.25}}{h} \qquad (9.1)$$

It may be concluded that this formulation is similar to the scour formulations in the ultimate state of Ramirez et al. (1990); while the non-dimensional relationship F_b is close to unity. In this case $h = 0.475q^{0.5}H^{0.25}$ is slightly greater than the effective cushion and determines the requirements of plunge pool depth, according to the falling energy and the incident flow. It must be noted that h depends more intensely on the specific flow.

If Ervine et al.´s (1997) equation is expressed in the general form of a theoretical circular jet, one finds that:

$$\frac{\Delta p_{max}}{D} = 29.27(1-C)L^+F_e \qquad (10)$$

$$F_e = \left(\frac{D^{1.33}H^{0.67}}{h^2} \right) = \left(\frac{2}{\pi g} \right)^{0.67} \frac{q^{1.33}}{h^2} = 0.162 \frac{q^{1.33}}{h^2} \qquad (10.1)$$

where an aeration coefficient for the jet C and an non-dimensional relationship F_e are included, the same which quantifies the action of flow impingement and the fall energy, with the water cushion requirements. It can be seen that flow impingement as well as height of fall are much more intense than in the case of a plane jet; hence, so is the need for a water cushion that dampens the action.

In Figure 2 the theoretical relationship of plane and circular jets is shown; ratios (S/B) vs. L^+ for the limit scour formulations of Ramírez et al. (1990), Wu (1974) and Lopardo et. al (1987); and the ratios $(\Delta p_{max}/B)$ vs. L^+ of the mean dynamic pressures formulations of Puertas (1994) and Ervine et. al (1997), including different values of F_b, C and F_e.

The pressures for the theoretical circular jet constitute a higher envelope, up to a value of the jet entry velocity on the order of around 2.40 times the critical velocity ($L^+ = 2.40$). From this point, the envelope of maximum pressure corresponds to the theoretical plane jet.

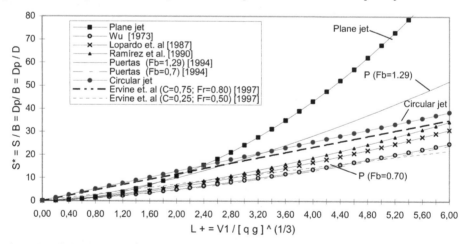

Figure 2. Relationship of mean dynamic pressure and ultimate scour formulation.

The extreme limits of the non-dimensional relationship $F_b = 1.29$ and 0.70 constitute higher and lower envelopes in pressures for the formulations of ultimate scour. This shows, as expected, that the maximum dynamic pressures are always reached when there is a minimum water cushion and vice versa. The minimum dynamic pressures will be obtained when the height of the water cushion is greater, coinciding in this case with Wu´s (1973) ultimate scour formulation; while with Lopardo et al.'s (1987) ultimate scour formulation, the approximation is carried out for an intermediate value of the above-mentioned non-dimensional relationship.

In Figures 3, 4 and 5, variations of the water cushion, total energy and energy dissipation are presented. These are calculated based on Moore's (1943) classic formula. The mean dynamic pressure for different values of F_b is according to Puertas (1994) and F_e is according to Ervine et. al (1997) and Ramírez et al.'s (1990) ultimate scour depth. The analysis was carried out for $q = 20$ m²/s and H = 25 - 250 m.

As for the depths of the water cushion (Fig. 3), it can be seen that the smallest depths of water cushion correspond to the contracted depth y_1, which theoretically ought always to contain the greatest amount of energy (or produce the least energy dissipation). Logically, the depth y_1 decreases as the height of the fall increases. However, it should be noticed that the real water cushion for the case of direct impact corresponds to the pool depth y_p; the same value is slightly greater than the effective cushion calculated from Puertas' formula, while it becomes the same from H = 170 m. It should be noticed that the pool depth does not completely surround the jet and, thus, there is no energy dissipation by jet diffusion.

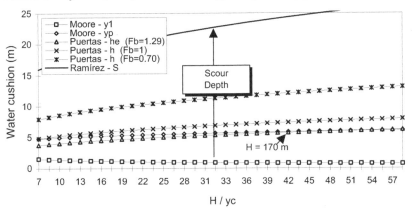

Figure 3. Water cushion depth: $q = 20$ m²/s; $y_c = 3.442$ m; H = 25 – 250 m.

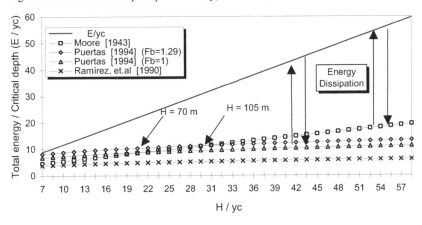

Figure 4. Energy dissipation: $q = 20$ m²/s; $y_c = 3.442$ m; H = 25 – 250 m.

In Figures 4 and 5 it can be seen that with Moore's formulation higher levels of energy dissipation are reached than those obtained with Puertas, up to H = 70 m for $F_b = 1.29$ and H =105 m for $F_b = 1$. This implies that, for the analysed flow, part of the jets would reach the floor in a more or less compact manner up to a height of 70 m, and then from this point an increase of air entrainment and break-up jet would be accentuated, resulting in an energy dissipation in the air which is not considered in Moore's formulation. In Puertas's formulation, since this phenomenon is implicitly registered in the tests, the above-mentioned loss is taken into account in some way.

Logically, the greater water cushion values are obtained with Ramírez et al.'s ultimate scour formulation and, thus, will always contain the least amount of energy (Fig. 4) or the production of greatest energy dissipation (Fig. 5).

The water cushion calculations obtained by Puertas' formulation increase proportionally with the reduction of the non-dimensional relationship F_b, containing less energy (Fig. 4) and producing a greater amount of energy dissipation (Fig. 5).

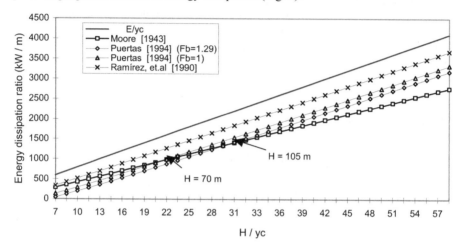

Figure 5. Energy dissipation relation: $q = 20 \ m^2/s$; $y_c = 3.442 \ m$; $H = 25 - 250 \ m$.

4 EROSIVE POWER OF THE IMPINGING JET

Since the choice of a typology for the dissipation of energy in a scour hole or on a rigid pool bottom will depend on the geological, geotechnical, economic and environmental conditions, it is interesting to find a direct relationship between the two design typologies, in practical terms of use. Thus, if it were known that a design in scour hole requires a scour height S, what equivalence would there be with a rigid bottom (or rock bottom) typology with a water cushion h? Which practical mechanisms should be considered in terms of pressure fluctuations and turbulence intensities? The answer to these questions would result in the choice of some type of design.

Following Annandale's work (1995), the "energy dissipation ratio" constitutes a parameter that reasonably represents the relative strength of the fluctuating disturbance and can be easily calculated. Thus, if the energy loss is ΔE and the specific flow q, the "energy dissipation ratio" per unit of width of flow is expressed as:

$$P = \gamma q \Delta E \qquad (11)$$

The relationship between "energy dissipation ratio" P and the material resistance can be expressed as the function:

$$P = f(K_h) \qquad (12)$$

at the resistance threshold. If $P > f(K_h)$, the resistance threshold is exceeded, and the material would be expected to fail. In our case, the required material constitutes a rigid plate or a rock, which resists to the pressure fluctuations that traverse the water cushion h. Thus, an alternative design for a basin with rigid bottom (plate or rock) and water cushion height h_p could be analysed with another basin and cushion height h_m. This, in turn, is related to the basic typology which constitutes the design of a scour hole with height S; the same height for the design conditions constitutes the basin with minimal resistance, since it has allowed the greatest energy dissipation ratio to develop. However, the difference in the energy production ratio DP between the scour hole P_s and the basin with a rigid bottom P_h, constitutes the net energy dissipation ratio "incremental energy dissipation" or, equivalently, the pressure fluctuations on the plate or rock bottom of the basin; thus:

$$DP = P_s - P_h \tag{13}$$

In Figure 6, the "incremental energy dissipation" (or the differences of energy production relation), can be found based on Ramírez et al.'s formulation. Moore and Puertas ($F_b = 1.29$, $F_b = 1.00$ y $F_b = 0.70$) formulations are analysed and it can be seen that the rigid plate design with Moore's criteria would be the least stressed for H < 45 m. However, it would be the most stressed for H > 105 m. In the other cases, behaviour is as expected: the most stressed situation corresponds to the design criteria with effective water cushions ($F_b = 1.29$).

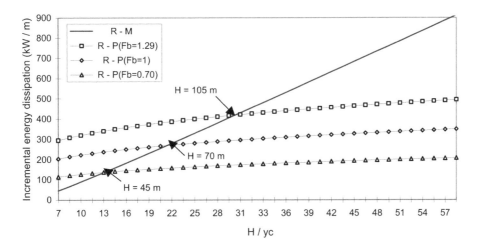

Figure 6. Incremental energy dissipation DP: q = 20 m²/s; y_c = 3.442 m; H = 25 – 250 m.

5 CONCLUSIONS

A combined analysis is presented of the ultimate scour and the mean dynamic pressure formulation. It is shown that they correspond to the same type of formulation. Different parametric analyses have been carried out, according to the height of fall.

A practical parameter is presented to estimate the resistance of the rigid pool bottom that is necessary to withstand the power of the water jet. It is called the "incremental energy dissipation", and allows the analysis of the different actions in the plunge pool. In this sense, a study should be carried out of the correlation between energy dissipation and bottom resistance, perhaps following the guidelines as outlined by Annandale, for the erosion ability phenomena.

It would be important to complete the research with the wall flow on the bottom and the pressures for water cushions that are ineffective.

REFERENCES

Aki, S. 1969. Jiyu rakka suimayaku no mizu-kusshon koka ni kansuru kenkyu (Estudio de la eficacia de los colchones de agua en vertido libre). Journal of the Central Electrice Research Institute, (Traducido al portugués por J.A. Pinto de Campos. LNEC, Lisboa).

Annandale, G.W. 1995. Erodobility. Journal of Hydraulic Research. IAHR, Vol. 33, NO. 4, pp. 471-494.

Armengou, J. 1991. Vertido libre por coronación en presas bóveda. Análisis del campo de presiones en el cuenco amortiguador. PhD. Thesis. Universitat Politècnica de Catalunya. Barcelona.

Beltaos, S. 1976. Oblique impingement of plane turbulent jets. Journal of the Hydraulics Division. Proceedings ASCE, Vol. 102, HY9 Sep.

Castillo-E, L.G.; Puertas, J. & Dolz, J. 1999. Discussion about pressure fluctuation on plunge pool floors. Journal of Hydraulic Research. IAHR, Vol. 37, N° 2, pp. 272-277.

Castillo-E, L.G. 1998. Revisión de las formulaciones de presión en los disipadores de energía en presas bóveda y corrección del coeficiente de presión dinámica. Comunicación personal. No publicado.

Castillo-E, L.G.; Puertas, J. & Dolz, J. 1996. Análisis conjunto de las formulaciones de socavación y presión dinámica media al pié de presas. XVII Congreso Latinoamericano de Hidráulica, IAHR. Vol 5, pp 179-190. Guayaquil, Ecuador.

Castillo-E, L.G.; Dolz, J & Polo, J. 1991. Acquisition and analysis of data to characterize dynamic actions in hydraulic energy dissipators. XXIV IAHR Congress. Vol. D, pp D-273 - D-280. Madrid.

Castillo-E, L.G. 1990. Comprobación y unificación de las formulaciones de la presión dinámica media de un chorro incidente en el punto de estancamiento y la zona de influencia. XIV Congreso Latinoamericano de Hidráulica. Vol. 1, pp 383-392. Montevideo, Uruguay.

Castillo-E, L.G. 1989. Metodología experimental y numérica para la caracterización del campo de presiones en los disipadores de energía hidráulica. Aplicación al vertido en presas bóveda. PhD Thesis. Universitat Politècnica de Catalunya. Barcelona, España.

Cola, R. 1966. Diffusione di un getto piano verticale in un bacino d´acqua d´altezza limitata. L´Energia Elettrica - N.11, pp. 649-667.

Cui Guang, T.; Lin, Y.; Liang, R. 1985. Efeito do impacto, no leito do rio, da lamina descarregada sobre uma barragem abobada. (Traducción del chino por J.A. Pinto de Campos). LNEC, Lisboa.

Ervine, D.A.; Falvey, H.T. & Withers, W. 1997. Pressure fluctuations on plunge pool floors. Journal of Hydraulic Research. IAHR, Vol. 35, N° 2, pp. 257-279.

Ervine, D.A. & Falvey, H.T. 1987. Behavior of turbulent water jets in the atmosphere and in plunge pools. Proc. Instn. Civ. Engrs., Part 2, 83, p.295-314.

Hartung, F. & Hausler, E. 1973. Scours, stilling basins and downstream protection under free overall jets at dams. Trans. of the 11th Congress on Large Dams. Madrid, Spain.

Horeni, P. (1956. Disintegration of a free jet of water in air. Byzkumny ustav vodohospodarsky prace a studie, Sesit p3, Praha, Pokbaba.

Lencastre, A. 1961. Desacarregadores de lâmina livre: bases para o seu estudo e dimensionamento. LNEC, Memoria No. 174. Lisboa.

Lopardo, R.A; Vernet, G.F & Chividini, M.F. 1987. Discussion to MASON, P.J. and ARUMUGAM, K. 1985. Journal of Hydraulic Engineering, ASCE, Vol.113, NO.9, Sep.

Mason, P.J.; Arumugam, K. 1985. Free jet scour below dams and flip buckets. Journal of Hydraulic Engineer, 111, No. 2, 220-235.

Moore, W.L. 1943. Energy loss at the base of a free overfall. Transactions, ASCE, Vol. 108, p.1343, with discussion by by M.P. White, p. 1361, H. Rouse, p. 1381, and others.

Muñoz, R. 1964. Socavación al pié de un vertedero. II Congreso Latinoamericano de Hidráulica. Vol. 2. Porto Alegre, Brasil.

Puertas, J. 1994. Criterios hidráulicos para el diseño de cuencos de disipación de energía en presas bóveda con vertido libre por coronación. PhD Thesis. Universitat Politècnica de Catalunya. Barcelona, España.

Ramírez, M., Fuentes, R. y Aguirre, J. 1990. Comparación unificada de fórmulas para la socavación al pié de presas. XIV Congreso Latinoamericano de Hidráulica. Vol. 3, pp 1467-1478. Montevideo, Uruguay.

Rajaratnam, N.; Aderibigbe, O.; Pochylko, D. (1995). Erosion of sand by oblique plane water jets. Proceedings Civil Institutions Engineers of Water, Maritime & Energy. 112, Mar. pp. 31-38.

Rajaratnam, N. 1981. Erosion by plane turbulent jets. Journal of Hydraulic Research. IAHR, 19, N0.4.

Schoklitsch, A. 1932. Kolbindung unter Überfallstrahlen. Die Wasserwirtsschaft.

Tennekes, H. & Lumley, J.L. 1972. A first course in turbulence. MIT Press.

Veronese, A. 1937. Erosioni di fondo a valle di uno scarico. Annali dei Lavori Publici. Roma, Italia.

Withers, W. 1991. Pressure fluctuations in the plunge pool of an Impinging Jet Spillway. PhD Thesis, University of Glasgow, February.

Wu, C.M. 1973. Scour at downstream end of dams in Taiwan. International Symposium on River Mechanics. Vol. 1. A13 1-6- Bankok, Tailandia.

Xu-Do-M. & Yu-Chang, Z. 1983. Pressao no fundo de um canal devido ao choque de um jacto plano, e suas características de flutuaçao. (Traducción del chino por J.A. Pinto de Campos). LNEC. Lisboa.

Pressure fields due to the impingement of free falling jets on a riverbed

J. Puertas
Civil Engineering School. Universidade Da Coruña

J. Dolz
Civil Engineering School. Universitat Politècnica de Catalunya

ABSTRACT: Free overflow spillways are commonly used in arch dams. The jet gains velocity as it falls, and it can generate a big impact over the rock downstream the dam. To avoid this, plunge pools are usually constructed in order to absorb a certain amount of the energy and to spread the jet. The more compact the jet is, the more severe is the impact it can generate. Some experiments have been done in order to evaluate the pressure fields over the plugeplunge pool floor. The energy dissipation due to the flight in the atmosphere has not been taken into account in most of these tests. This paper presents some results about the pressure field one can expect in these effects are considered.

1 INTRODUCTION

Because of the geometry of arch dams, free overflow spillways are the most common ones. The jet that spills from the crest of the dam falls without any external action, till it reaches the base of the dam. Obviously, the water gains velocity as it losses its elevation, and when it reaches the base it has a huge amount of kinetic energy, which means strong actions over the rock (or the reinforced substrate) downstream the dam.

The loss of energy of the jet within the atmosphere, and the modifications of its shape because of the effect of the air entrance, has been pointed out by Ervine and Falvey (1997).

To avoid a direct contact between the jet and the subgrade it is usual to construct an auxiliary structure (downstream cofferdam), which creates an upstream water pool between the downstream wall of the dam and its upstream wall.

If we consider the slab in the bottom of the pool (a natural rock slab or a concrete slab), the impingement of a jet and its momentum variation causes a pressure increment. The impact point is called stagnation point. This overpressure is a dynamic phenomenon. In addition, as the slab has a cushion of water over it, there is a hydrostatic term of pressure. In this paper, only the dynamic term of the pressure will be considered.

From the sixties on, the pressure field in the plunge pool has been studied. A classical reference is the one by Cola (1966), that proposes in a theoretical way (applying the momentum conservation law) formulae to evaluate the mean dynamic pressure in or near the stagnation point. The parameters involved and the structure of the equation were:

$$\Delta P = C\rho \frac{v^2 B}{2h} e^{-K\left(\frac{x}{h}\right)^2} \tag{1}$$

where C and K are coefficients to be fit by experimental ways, v is the velocity of the water when reaching the pool, B is the thickness of the jet, h is the thickness of the cushion and x is the distance to the stagnation point. Cola made some hypothesis, including the absence of air entrance and a Gaussian pressure distribution.

Some researchers have carried out tests that give an order of magnitude of the parameters (C, K):

Table 1: Values of the parameters of Cola's formulation

Author	C	K	Observations
Cola (1966)	7.18	40.51	
Hartung&Hausler (1975)	5	19.6	Data obtained from the reference
Hartung&Hausler (1975)	3.56	9.92	Data modified by the authors, according with spreading angles obtained from Ervine and Falvey (1987)
Franzetti (1980)	2.15-5.74		

In the table above, only the results by Hartung & Hausler modified by the authors and the results from Franzetti take into account the effect of the flight. The results from Franzetti have been obtained from a scale model of an arch dam. The results by Hartung & Hausler are based on a theoretical approach, and they incorporate a spreading angle that depends on the turbulence of the jet, which has been defined by Ervine and Falvey (1987). As it can be seen, those results are quite lower than the others, which means that the effect of the flight cannot be neglected. This is clearly pointed out in Ervine et al (1997). On this paper, further information about this topic will be provided.

2 EXPERIMENTAL METHODOLOGY

2.1 Experimental facilities

The experimental work, which is the basis of this paper, has been fully developed in the Hydraulics Laboratory of the Universitat Politècnica de Catalunya (Barcelona, Spain). The equipment used in this research was:

- A structure based on two orthoedric bowls, where the upper one spills the water to the lower one, in which the cushion is formed and pressures are measured. The spillway (1.2 m wide) is straight, in order to work with bi-dimensional conditions. The vertical distance between the bowls may be varied in order to work with different falling heights, from 2m up to 5.5m. The thickness of the cushion and the discharge may also be varied: 0<h<0.8m, 0.02 m²/s<q<0.2 m²/s (Figure 1).

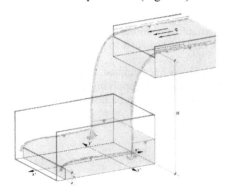

Figure 1. - Experimental facilities. Sketch of the experimental structure.

- A set of piezo-resistive sensors, which measure with a frequency of about 40 Hz during one minute (2200-2400 measurements each test). Each test includes measurements in a range of 4 to 6 different locations in 6 cm steps, including the stagnation line and some positions both upstream and downstream.

Figure 2. - Set of piezo-resistive sensors located at the bottom of the lower bowl.

The experimental work consisted of 184 tests, with 5 different heights, and a wide variation of discharges and cushion thickness. The pressure fields recorded in these tests were over 1000.

2.2 Dimensional analysis

In order to apply the results of this research to all structures including a falling nappe, we used non-dimensional parameters. The parameters involved (after the application of Buckingham's theorem, taking into account pressure, gravity and inertia forces) are:

- $N1 = \dfrac{p}{\gamma H}$ (mean dynamic pressure recorded versus falling height)

- $N2 = \dfrac{h}{H}$ (cushion thickness versus falling height)

- $N3 = \dfrac{q}{\sqrt{2gH^3}} = \dfrac{B}{H}$ (nappe thickness versus falling height; Froude number)

- $N4 = \dfrac{x}{H}$ (distance to the stagnation point versus falling height)

where q is the discharge per unit width, and H is the falling height measured from the energy line in the upper bowl till the water surface in the lower bowl

The main topics to be discussed in this paper are the relationship between N1 and N3 (pressure versus discharge, at a given depth), and the relationship between N1 and N4, as it is different from 0 (as the point recorded is not over the stagnation line).

3 RESULTS

3.1 Introductory remarks

3.1.1. Effective cushion and compactness of the nappe

The behavior of the cushion when a nappe is impinging depends on its thickness. The frontier between effective and non-effective cushions proposed by Cui Guang Tao el al. (1986) depends on the relationship B/h, where h is the thickness of the cushion and B is the thickness of the nappe as it enters the cushion.

This relationship does not take into account that both the height of the falling nappe and the discharge promotes the energy of the cushion. The only parameter B cannot explain this phenomenon. The parameters involved in the creation of a hydraulic jump into the pool are, on the one hand, H and q: the higher they are, the more energy the nappe will have; and on the other hand, the thickness of the cushion. Thus, the parameter that could explain the frontier must be based on these parameters.

The non-dimensional number N5, defined as $N2^2/N3$ gives a relationship between forces trying to eject or to maintain the cushion. This number accomplish with all the conditions expected about parameters involved (q, h, H). As it is detailed in Puertas (1994), there is a clear relationship between the values of N5 and the existence of an effective cushion. The areas with and without an effective cushion are separated by the frontier N5=0.6. If N5 has a value of over 0.6, there is an effective cushion.

As mentioned before, the mean dynamic pressure, Cola (1966), can be written as:

$$\Delta P = C\rho \frac{v^2 B}{2h} \quad \text{or} \quad N1 = C \frac{N3}{N2} \quad (N4=0) \tag{2}$$

This expression includes the term h (N2), that is to say, the thickness of the cushion. If the cushion is not effective, h is not a relevant term and the expression by Cola would be inappropriate.

Thus, two behaviors may be observed, one for each set of data. The data from effective cushions can be fit quite well to Cola's expression, while the data from non-effective cushions do not have a good adjustment. This may be explained by the fact that h has no influence and so the expression by Cola is not suitable for this set of data.

The compactness of the jet when reaching the slab, which has been evaluated in terms of h/B (Hartung and Hausler, 1973), can be appreciated by comparing the values of N1 vs. N2/N3 (h/B). From the results of our tests, it can be seen that the full range of the tests seems to cover a region in which the core of the jet has been altered, as the values of N1 do not reach the value of 1, and there is not a zone in which the values of N1 (dimensionless dynamic pressure) are independent of the values of N2/N3, as in the tests by Ervine et al (1997).

3.1.2. Breaking of the nappe

If the cushion is not effective, the effect of h is considered to be negligible, so a dimensional analysis provides an expression in which the discharge has no influence and its meaning (quite evident) is the direct transformation of the kinetic energy (which corresponds with the falling height) into pressure over the slab. The only way of losing energy is now from the flight into the atmosphere: the studies by Ervine et al. (1997) give us both a physical explanation and a quantitative formula to evaluate if the jet has already broken its structure when reaching the slab. With the data available from our experiment, a relationship between N1 (5%) (mean dynamic pressure only achieved by a 5% of the data in one test, this figure is considered as the maximum of the test) and N3 (discharge) can be obtained for data from non-effective cushions. A value for N1 of close to 1 can be expected for compact nappes, and it will be smaller when the degree of compactness of the nappe decreases (because of large falling heights or small discharges –small values for N3, in all cases-).

The maximum values are observed instead of the mean values because of the vibration of the nappe, that produces some variations in the location of the stagnation line, which can be noticed as a decrease of the mean values, as the pressure sensor is located in a fixed position, while the nappe has some variations in the position of the impact. The maximum values can be considered as representative, but the mean values incorporate measurements with no impact data.

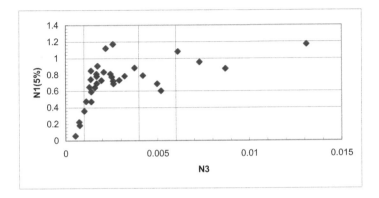

Figure 3. Reduction of the nappe energy due to its flight (N1 (5%)-Maximum dimensionless pressures-vs. N3-discharge).

As can be observed in figure 3, high N3 (Froude number, dimensionless discharge) values result in values for N1 (pressure) of close to 1. For values between 0.0011 and 0.005 the effect of the flight reduces the amount of energy some 20%, and for values under 0.0011 the decrease of N1 is quite evident. Therefore, it would seem that N3=0.0011 could be a frontier (this must be confirmed with further research), to determine whether the nappe remains as a continuous structure or simply as a set of clumps.

As an example, two single tests will be analysed. For the first one, N5=0.2 and for the second one N5=10: that is to say, the first test corresponds with a non-effective cushion and the second one with an effective cushion.

Table 2. Hydraulic parameters of the tests analysed.

Test	Q (l/s)	H+h (m)	h(m)	N2	N3	N5
1	71.63	5.45	0.08 (n.e.)	0.015	0.00108	0.2
2	71.63	5.45	0.56 (e.)	0.11	0.0012	10

These two tests are both within the N3 border area, but their behaviour are quite different, as it will be seen in the histograms showed below, which presents the distribution of the registered pressures in the stagnation point during the tests (2001 points each histogram).

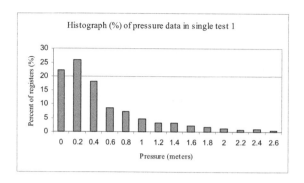

Figure 4. Histogram of pressure data for a non-effective cushion test.

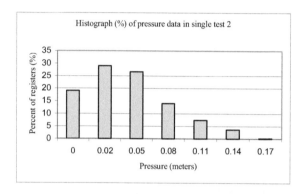

Figure 5. Histogram of pressure data for an effective cushion test.

Some conclusions can be obtained from figures 4 and 5: as N3 values are quite low, no values near N1=1 (pressures about 5 m) are registered in any case. The nappe seems to be broken when it enters the cushion.

In the test shown in figure 4 there is no effective cushion, so maximum pressures are about 2 m (H1=0.4). A big amount of zero data is registered, and also big amounts of nearly zero data. This is because of the oscillation of the nappe, which does not impact over the same point during the full time of the test. The histogram is quite wide; a mean value has not a statistical sense.

In the test shown in figure 5, a clear mean value is obtained around 0.044 m, and the range of registered pressures is very narrow. No values over 0.2 m are registered, so the cushion has a huge dissipation effect. The oscillation of the nappe is also dissipated along the cushion, so the values registered are quite stable. Zero values correspond with no dynamic effects. Anyway, the maximum values of N1 are about 0.04 (virtually nil).

3.2. *Pressure distribution*

The experimental program did not include but a few amount of tests with N3<0.0011. If this data are removed, and the equation 1 is fit for the tests with an effective cushion, a coefficient of 3.8 is obtained.

$$\frac{\Delta P_{max}}{H\gamma} = N1 = 3.8 \frac{N3}{N2} \tag{3}$$

This value is much lower to those obtained by Cola and Hartung & Hausler (table 1), but it is quite similar to the one assuming a fully turbulent jet from the data by Hartung and Hausler and also to the data by Franzetti (table 1).

The decay of the mean dynamic pressures when leaving the stagnation line is expressed by Cola as a Gaussian distribution:

$$\Delta P = \Delta P_{max} e^{-K\left(\frac{x}{h}\right)^2} \tag{4}$$

The fitting of the parameter *(K)* of this distribution is considered to be difficult (Lencastre (1961), Armengou (1991)). When analyzing the experimental data, the distribution may be assumed to be more or less symmetric with the stagnation line as the axis, but the shape of the distribution do not seem to fit a Gaussian formulae. So, a possible distribution could be:

$$\frac{\Delta P}{\Delta P_{max}} = e^{-K\left(\frac{x}{h}\right)^\alpha} \tag{5}$$

If the exponent equals 2 this formula is the one proposed by Cola. A set of individual fittings of K parameter has been done with every set of points belonging to an individual test (including from 4 to 6 points) with effective cushion. More than 100 fittings have been calculated. As 4 or 6 points are a very few amount of information to fit two parameters, the exponents have been considered as a constant, and only the values of K have been fit. The values for the exponent have been fixed as 2, 1 and 0.5. In table 1, the mean values of K and their standard deviation are shown.

Table 1. Pressure distribution. Values of K.

a	Mean K	ST. Dev.K	Coef. of variation
2	8.67	14.15	1.63
1	3.02	2.04	0.67
0.5	1.98	1.01	0.51

The parameter K corresponding to a=2 (8.67) is similar to the one obtained from the data by Hartung & Hausler modified as proposed by Ervine & Falvey for fully turbulent jets. Anyway, if considering the coefficient of variation (standard variation/mean) as a quality parameter, it is quite evident that the fitting is much better when the exponent (a) is fixed as 0.5, while the fitting with a Gaussian function is quite poor.

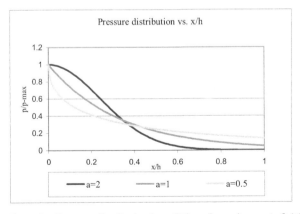

Figure 6.   Pressure distribution best fittings for a given set of (a) exponents.

The analysis of the data explained above considers every individual experiment to be fit to a curve. As the variables involved are expressed in a non-dimensional way, a single fitting can be done to adjust the full range of measurements, in spite of the fact that the measurements belong to different experiments.

As expressed above, a parameter fitting can be proposed in order to adjust the value of the exponent. The equation to be fit has been written as:

$$\ln(\frac{\Delta P}{\Delta P_{max}}) = A - K\left|\frac{x}{h}\right|^{\alpha} \qquad (6)$$

where the value of A is expected to be 0. The hypothesis A=0 with a confidence range of a 90% is considered as a quality control.

Table 2. Pressure distribution. Values of K.

α	A	K	R^2	¿A=0? 90%
1	-0.21	2.6	48	
2	-0.7	1.5	28	
0.5	0.2	2.66	50	A=0
0.4	0.263	2.55	47	
0.6	0.134	2.73	51	A=0
0.3	0.279	2.3	42	
0.7	0.047	2.75	51	A=0

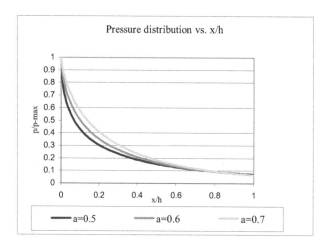

Figure 7. Values of (a) which give the best fitting.

The first goal is to achieve a good result for test A=0 (by a "t" test, with a 90% reliability), so this parameter must equal zero if the measurements correspond to the stagnation point. The best values for the "R^2" parameters are to be chosen from all the exponents that accomplish A=0.

On analyzing Table 2, in which only the data with effective cushion have been included, best values for (a) are seen to be 0.5 and 0.6 or 0.7 (not for 2, in any case). It may therefore be accepted that the exponent is about 0.5. In addition, the K-values are quite similar to those obtained in table 1 when the values of (a) are 0.5 and 1, while the values of K when a=2 are very different, which means that the reliability of the fitting is very poor. Anyway, the R^2 values are not high in any case, so this phenomenon must need further research, and it can be considered to have a high degree of random behavior.

If comparing the shapes of the curves in figure 5, we can see that the lower the value of a, the more narrow the area with high values (i.e. >0.6) of the ratio of dynamic pressures is. The best values of the exponent seem to correspond with quite low values, that is to say, with a narrow line of high dynamic pressures. That makes a difference with the analysis made by Cola and others, which states that the best fitting is the Gaussian distribution, which gives a high pressure area quite wide.

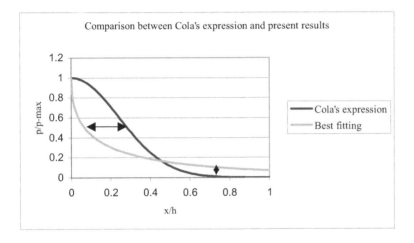

Figure 8. Comparison between Cola's expression and present data fitting.

This is probably the main topic of this paper, as this difference can be explained by the lack of compactness of the nappe, which reduces the energy of the jet and also the width of the high pressures area.

The pressure distribution obtained from our data shows two differences with Cola's theory (figure 8): the area with high pressures is narrower than the one proposed by Cola, and the area with noticeable pressures (not high, but clearly registered) is quite wide, which means that the jet suffers a diffusion and a decrease of energy much bigger than the one a compact jet suffers. The aeration of the nappe during its flight and the existence of an effective cushion favor this behavior. So it is quite reasonable that our data show a pressure distribution not as compact as Cola's.

So, an expression that can be proposed, as a first approach, is:

$$\frac{\Delta P}{\Delta P_{max}} = e^{-2.66\left|\frac{x}{h}\right|^{0.5}} \tag{7}$$

4 CONCLUSIONS

The effect of the flight in the atmosphere over the nappe can be noticed as a lack of compactness, which generates a pressure field with values clearly lower than the ones expected. The distribution of pressures shows a wide area of noticeable pressures but a very narrow area of high pressures, due to the spreading of the jet. The effect of a compact jet would be much more strong, so the flight effects clearly collaborate to the energy dissipation.

5 REFERENCES

Cola, R.; Diffusione di un getto piano verticale in un bacino d'aqua d'altezza limitata; *L'Energia Elettrica-* n.11 -1966. pp. 649-667

Cui Guang Tao, Lin Ji Yong, Liang Xing Rong; Effect of the impact of a jet over the riverbed; *Shuili xuebao*, Beijing, (8), (translation (Portuguese) J.A. Pinto de Campos, LNEC, Lisboa) 1985, pp.58-63)

Ervine, D.A., Falvey, H.T.; Behavior of Turbulent Water Jets in the Atmosphere and in Plunge Pools; *Proc. Inst. Civ. Engrs.*; Part 2, 83, Mar. 1987, pp.295-314

Ervine, D.A.; Falvey, H.T., Withers, W.; Pressure fluctuations on plunge pool floors; *Journal of Hydraulics Research*, vol 35, n.2, 1997

Franzetti, S.; Pressione idrodinamichi sul fondo di una vasca di smorzamento; *L'Energia Elletrica,* N.6, 1980, pp.280-285

Hartung, F., Hausler, E.; Scours, Stilling Basins and Downstream Protection Under Free Overfall Jets and Dams; *Trans. of the 11th Congress on Large Dams*, Madrid, Spain, Q.41, R.3.

Lencastre, A.; Descarregadores de lâmina livre: bases para o seu estudo e dimensionamento; *Laboratorio Nacional de Engenharia Civil*, Memoria No. 174. Lisboa, 1961.

Puertas, J; Criterios hidráulicos para el diseño de cuencos de disipación de energía en presas con vertido libre por coronación. *Doctoral Thesis*, Universitat Politècnica de Catalunya, 1994

Jet aeration and air entrainment in plunge pools

Gebidem Arch Dam (Switzerland)
(Courtesy of Electra-Massa Company)

Does an aerated water jet reduce plunge pool scour?

H.-E. Minor, W.H. Hager & S. Canepa
VAW, ETH-Zentrum, Zürich, Switzerland

ABSTRACT: The effect of the air concentration of a water jet on plunge pool scour was investigated by using a simplified experimental configuration. Instead of considering the complete arrangement involving chute and bottom aerators to create an aerated jet, a pre-aerated pressurized flow was produced with a pipe. Discharge of water and air was varied and measured so that air concentration and jet diameter close to impact on the free water surface were known. Three different sediments were tested. The results show clearly that the scour depth reduces with increasing air concentration although scatter is large.

1 INTRODUCTION

Energy dissipation at large dams with a significant design discharge often is accomplished by trajectory spillways in combination with a plunge pool which normally proves more economic than a stilling basin. The spillway / plunge pool solution may be considered if geological conditions are adequate and the developing scour is acceptable. The scour development is a significant concern because undermining may endanger the dam foundation and adjacent hydraulic structures including the flip bucket. Steep valley sides may become unstable because their foot is eroded or the spray is moistening the overburden material, thus creating landslides. Bars formed by these slides or by aggradation downstream of the scour may block the water flow producing backwater on the powerhouse, the bottom or irrigation outlets. Additionally, the spray may have adverse effects on installations like the switchyard.

The main question with respect to safety, however, is the size of the plunge pool scour including especially the scour depth. Today, it is common practice to introduce bottom aerators in high head spillways to reduce cavitation risk. The air concentration of the jet leaving the flip bucket is much higher in these cases. It is governed by the number of aerators applied and the distance of the last aerator from the flip bucket.

Figure 1 shows a photo of Restitucion trajectory spillway which bypasses the powerstation and comes into operation when the powerstation is shut down and Mantaro powerstation situated above continues operation. With a hydraulic head of 248 m and a maximum discharge of 90 m^3/s, the spillway was designed such that the scour for all discharges develops in the river bed. As is clearly visible from Figure 1 the jet is aerated from the bottom by aerators in addition to the natural aeration from the water surface thus increasing the air concentration considerably. The jet remains remarkably compact although its thickness increases significantly along the trajectory.

The question is whether an additional jet aeration has an influence on the scour depth in the river bed. Unfortunately, not much information is presently available. A general review on plunge pool scour was presented by Whittaker and Schleiss (1984). More recently, Mason (1984, 1989) and Mason and Arumugam (1985) have added valuable information. Mason (1989) also conducted hydraulic experiments with a rectangular pre-aerated water jet that was submerged by the tailwater. In the following, a different setup will be described with results of a recent research project being presented (Canepa 2002).

Figure 1. Typical trajectory spillway with a significantly aerated jet (Photo: G. Soubrier)

2 EXPERIMENTS

Hydraulic modeling of a trajectory spillway including free jet flow in the air and scour at its impact location needs large laboratory space, provided that the model is sufficiently large to inhibit scale effects. Also, definition of the air-water mixture jet just upstream of impact onto the water body in terms of average velocity, mean air concentration and jet area requires some simplification in the model. In order to control the jet impact conditions, a simplified jet arrangement was considered in the present project, consisting of a free pre-aerated jet issued from a circular pipe shortly upstream from impact onto the water surface covering a sediment bed. The air-water jet was generated by a plexiglass pipe of diameter D=0.10 m, and a hose of diameter D=0.019 m to check the effect of jet thickness. Air was supplied to the pipe up to a discharge ratio $\beta=Q_a/Q=3$, where Q_a is air discharge and Q is water discharge (Fig. 2).

The plunge pool was simulated with a sediment corresponding to a completely disintegrated rock surface that may establish after a sufficiently long stress. Three crushed almost uniform stone chip sediments were employed whose typical sizes were d_{90}=0.056 m, d_{90}=0.019 m and d_{90}=0.0096 m, with a standard deviation $\sigma=(d_{84}/d_{16})^{1/2}$ smaller than 1.40. The sediments were set with a thickness of approximately 0.30 m in a rectangular channel of 0.70 m depth, such that another 0.30 m of water depth could be supplied. The sediment was inserted horizontally prior to each experiment with a grading mechanism. In total, 38 experiments were conducted, of which 17 were water experiments, and 21 air-water mixture tests.

Experiments started once the sediment surface and the water elevation were adjusted, and air was discharged into the plexi-glass pipe. Then, water was added, and time t set to zero. Both the sediment and water surface profiles were measured once end scour conditions had established (Fig. 3). The sediment surface profile tended quickly to an end scour profile, especially for the large sediment size. The sediment profile was also measured at intermediate times to determine the advance of scour with time; these results are not presented here, however. Significant differences were observed between wet and 'dry' scour conditions once water was drained from the channel, the latter producing always less scour depth. Lowering the water level in the channel and within the sediment led to rearrangement and small slides in the scour hole. In the following the focus lies on maximum scour depth and the influence of jet air concentration on scour depth.

118

Figure 2. Experimental Setup

3 EXPERIMENTAL RESULTS

3.1 Basic scour parameter

Figure 3 shows a definition plot for plunge pool scour as investigated in the experiments described. An air-water jet of water discharge Q and air discharge Q_a having a diameter D and an angle α relative to the horizontal was subjected onto a water body of initial depth h_o above a horizontal sediment layer of grain size d_{90}. Once the sediment surface developed into a close to end scour condition, it may be characterized with the maximum scour depth z_m at location x_m relative to the scour origin at x_o. The maximum aggradation height z_M and the maximum surface scour width b_M are at locations x_M (Fig. 3).

The many relations proposed to predict scour depth caused by a plunging jet differ greatly. Different parameters are considered important by the different researchers. Almost all of them consider the influence of h_o important so that z_m is often directly combined with h_o such as $z_m + h_o$ or z_m / h_o. This could not be substantiated for the data of the present study. The other main parameters considered important in the literature are specific discharge q, different definitions of hydraulic head, jet thickness, angle of impact and sediment size.

It has proven advantageous to use the Froude number as dominating parameter. In the following two Froude numbers will be used. The Froude number $F_j = V/(gD_w)^{1/2}$ describes the jet issued by the pipe, with the velocity of the water and the air-water-mixture, whereas D_w is the jet diameter. For clear water jets D_w is equal to the pipe diameter. For air-water mixture jets a computational diameter D_w was used, however, corresponding to the area the water flow only.

It must be kept in mind that in an air-water mixture jet the velocities of the water particles and the air in the jet are equal, and the computational area occupied by the water alone is smaller than the diameter of the mixture flow which corresponds to the pipe diameter.

Usual scour hydraulics may be characterized with the densimetric particle Froude number $F_d = V/(g'd_i)^{1/2}$ where V is jet velocity, $g' = [(\rho_s - \rho)/\rho]g$ reduced gravitational acceleration, and $d_i = d_{90}$ is the determining sediment size. It has proven advantageous to use d_{90} as determining sediment size, as many other authors do. This definition of Froude number combines the characteristics of jet flow with sediment characteristics.

3.2 Maximum scour depth

Figure 4 shows the maximum scour depth z_m versus the jet Froude number F_j for the three different sediments used. The influence of air concentration on maximum scour depth is obvious.

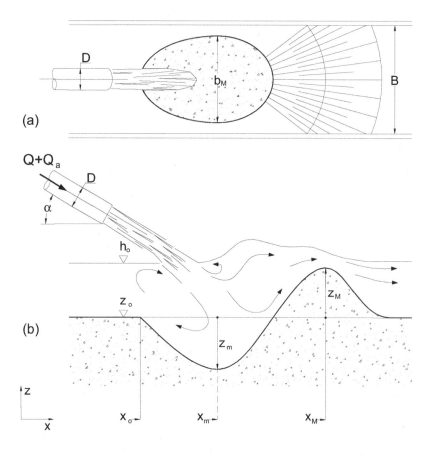

Figure 3. Definition sketch for plunge pool scour (a) plan, (b) longitudinal section

Because of the limited amount of data a definite dependence of the various parameters cannot be found by this way of presentation.

A remarkable result is achieved for clear water (subscript w) scour without aeration, when the dimensionless maximum scour depth $Z_w = z_w/D_w$ is plotted against densimetric particle Froude number F_d (Fig. 5). The densimentric Froude number varies between 2.5 and 20 for the three sediments involved, and the data follows reasonably well the linear relation

$$Z_w = 0.37\, F_d \tag{1}$$

Accordingly, the significance of the densimetric Froude number for the scouring process is experimentally verified. To detect the effect of air content on scour depth, white water data were plotted as z_m/z_w versus β. Using equation (1) $Z_w = 0.37\, F_d$ and remembering that $D \equiv D_w$ in these experiments, the result obtains

$$\frac{z_m}{z_w} = \frac{Z_m \cdot D}{Z_w \cdot D_w} = \frac{Z_m}{Z_w} = \frac{Z_m}{0.37\, F_d} \tag{2}$$

Figure 6 shows that the scatter of data is considerable. The points for clear water ($\beta = 0$) are also included and show a scatter of approximately 25 %, as all the other data. This is not so clearly visible in Figure 5.

Nevertheless, Figure 6 shows a tendency that maximum scour depth reduces with increasing air content of impinging jets. The data may be fitted by

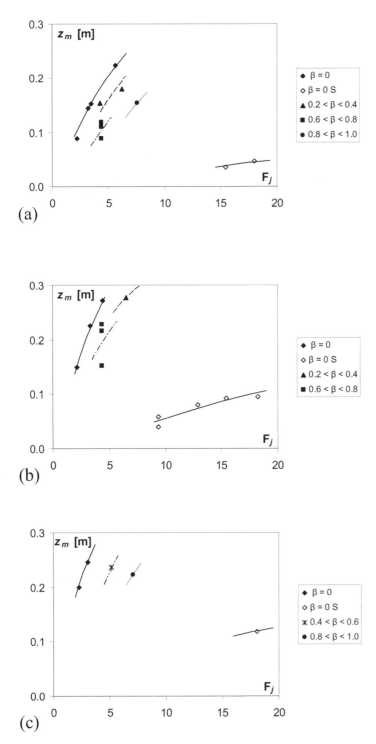

Figure 4. Maximum scour depth z_m versus jet Froude number F_j with air content β as third parameter. Grain size d_{90} = (a) 0.056 m, (b) 0.019 m, (c) 0.0096 m. Addition S means a jet diameter D=0.019 m, all other points refer to a jet diameter of D=0.10 m

121

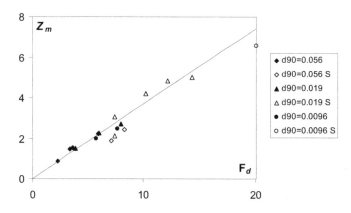

Figure 5. Normalized maximum scour depth versus densimetric particle Froude number F_d

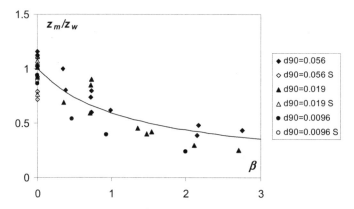

Figure 6. Effect of air content β on maximum scour depth

$$\frac{z_m}{z_w} = \frac{1}{(1+\beta)^{0.75}}$$ (3)

Figure 6 also indicates that already a small amount of air involves a significant reduction of scour depth.

PRACTICAL RESULTS

The results of the present study do not give a direct answer to the question: How deep will the prototype scour be for an aerated jet originating from a trajectory spillway. However, it is possible to predict at least approximately the effect of jet air concentration on scour depth for prototype conditions.

If the jet air concentration immediately upstream of the tailwater level is known its influence on the maximum scour depth compared to a jet without any air may be calculated with

$$z_m = \frac{z_w}{(1+\beta)^{0.75}}\tag{4}$$

The influence of the additional aeration may be estimated by applying (4) for two air concentrations β_1 and β_2 with the corresponding maximum scour depth $z_{m,1}$ and $z_{m,2}$

$$\frac{z_{m,2}}{z_{m,1}} = \frac{(1+\beta_2)^{0.75}}{(1+\beta_1)^{0.75}}\tag{5}$$

The procedure would be to calculate the maximum scour depth by applying one of the available relations, e.g. presented in Whittaker and Schleiss (1984) or by Mason (1984, 1989) taking into account of jet expansion (Ervine and Falvey 1987) and its aeration while travelling through the atmosphere. Then an estimate should be made how much higher the air concentration is expected to be at the point of impact, due to bottom aeration in the spillway chute. By applying equation (5), the reduction of maximum scour depth due to increased air content of the jet may be estimated.

SUMMARY

Plunge pool scour due to trajectory jets was investigated with a simplified model arrangement to improve the definition of hydraulic boundary conditions. Instead of using a full setup including spillway, flip bucket for a jet travelling through the atmosphere that eventually impinges on the loose downstream sediment bed, a pressurized air-water mixture was considered. Accordingly, the jet diameter and the jet air content prior to impact on the tailwater are accurately known and allow the analysis of plunge pool scour features. It was demonstrated that the maximum scour depth due to pure water jets varies linearly with the densimetric particle Froude number F_d. An increasing air concentration of the jet impinging on the tailwater and the underlying sediment reduces the maximum scour depth. More detailed investigations are necessary, however, to produce final design criteria.

REFERENCES

Canepa, S. (2002) Diploma work, Versuchsanstalt für Wasserbau, Hydrologie und Glaziologie, ETH Zürich, Supervisors: H.-E. Minor, W.H. Hager, unpublished.

Ervine, D.A. & Falvey, H.T. (1987) Behaviour of turbulent water jets in the atmosphere and in plunge pools. *Proc. Institution of Civil Engineers* 83: 295-314; 85: 359-363.

Mason, P.J. (1984) Erosion of plunge pools downstream of dams due to the action of free-trajectory jets. *Proc. Institution of Civil Engineers* Part 1 76: 523-537; 78: 991-999.

Mason, P.J. & Arumugam, K. (1985) Free jet scour below dams and flip buckets. *Journal of Hydraulic Engineering* 111(2): 220-235; 113(9): 1192-1205.

Mason, P.J. (1989) Effects of air entrainment on plunge pool scour. *Journal of Hydraulic Engineering* 115(3): 385-399; 117(2): 256-265.

Whittaker, J.G. & Schleiss, A. (1984) Scour related to energy dissipators for high head structures. *Mitteilung* 73. Versuchsanstalt für Wasserbau, Hydrologie und Glaziologie VAW, ETH Zurich.

NOTATION

b_M maximum scour width
D pipe diameter
D_w computational equivalent water jet diameter
d_{90} sediment size
h_o flow depth
F_d $= V/(g'd_{90})^{1/2}$ densimetric Froude number

F_j $=V(gD_w)^{1/2}$ jet Froude number
g   gravitational acceleration
g' $=[(\rho_s-\rho)/\rho]g$ normalized gravitational acceleration
q   discharge per unit width
Q   water discharge
Q_a air discharge
V   cross-sectional velocity
x   streamwise coordinate
x_m location of scour maximum
x_M maximum bar location
x_o scour beginning
z_m maximum scour depth
Z_m $=z_m/D_w$ relative maximum scour depth
z_M maximum bar height
z_w maximum scour depth for water jet
Z_w $=z_w/D_w$ relative maximum scour depth for water jet
α jet angle
β $= Q_a/Q$ jet air content
ρ water density
ρ_s sediment density
σ sediment non-uniformity

Reduction of plunge pool floor dynamic pressure due to jet air entrainment

J.F. Melo
LNEC - National Laboratory of Civil Engineering, Lisbon, Portugal

ABSTRACT: This paper addresses the favourable effect produced by air entrained by free-falling jets on the mean dynamic pressures on plunge pool floors. An analysis of the information available in literature is presented. A combination of independent proposals by different authors is compiled into an equation, allowing the estimation of the effect of aeration in the dynamic pressure on the pool floor. A comparison is made between results given by the suggested equation and experimental data obtained by the author with submerged rectangular and vertical air-water jets.

1 INTRODUCTION

Promoting lateral spread and disintegration of free-falling jets is generally accepted as advantageous in terms of energy dissipation. In fact, phenomena associated to lateral spread are directly connected to internal turbulence of jet flow, and therefore associated to the increase of kinetic energy dissipation along the jet atmospheric phase. Also, jet cross-section growth due to lateral spread decreases jet power per unit area at the impact section with the plunge pool free surface.

If jet disintegration is reached before impact with the pool, a considerable increase in free jet energy dissipation rate is obtained, mainly because atmosphere imposes a greater resistance to the movement of water droplets than to compact water bodies, such as undeveloped or partially developed free jets.

To promote lateral spread and aeration, thus favouring kinetic energy dissipation, the incorporation of splitters and deflectors have been considered in many overflow spillways.

Although it is generally accepted that air entrained into plunge pools produces an increase in the diffusion rate of submerged jets and, therefore, a reduction of dynamic action on plunge pool floors, it is recognized that there is still a lack of sufficiently detailed information concerning quantitative effects of the entrained air in the dynamic pressures.

Most studies of air entrainment by free-falling jets and its effects in submerged-jet diffusion flow were developed in the last 30 years, mostly with the purpose of obtaining a better understanding of the complex phenomena involved in two-phase highly turbulent flows and adequate formulae for its quantification.

The analysis presented in this paper was developed using information available in specialised literature and in a compilation made by Amelung (1995) and complemented by Melo (2001). Concerning the various research works in the specific topic of this paper, the following ones are worth mentioning: earlier works by Homma (1953), Cabelka (1955) and Kraatz (1965); subsequent works carried out by Ervine & Elsawy (1975), Rao & Kobus (1975), Ervine (1976), McKeogh (1978), Ervine et al. (1980), Falvey (1980) and Ervine & Falvey (1987); more recent research works by Armengou (1991), Amelung (1995), Ervine et al. (1997), Bohrer & Abt (1996), Melo (2001) and Bollaert & Schleiss (2001).

Data relating to the effect of air entrained by free jets into plunge pools, which are presented here, were obtained in an experimental facility developed at LNEC, Portugal, in which water and air flows were controlled independently, thus allowing the simulation of jets with similar power at impact, but with distinct imposed air concentrations.

2 AIR ENTRAINMENT AT FREE JET IMPACT SECTION WITH THE PLUNGE POOL

Concerning the mechanism of air entrainment into plunge pools, both Rao & Kobus (1975) and Ervine et al. (1980) consider that, when a free jet with significant velocity impacts the pool free surface, it entrains the surrounding air mass from the boundary shear layer and the jet surface irregularities and cavities. However, this explanation is merely qualitative.

Ervine et al. (1980) identify the existence of four different situations concerning the air entrainment of free jets into plunge pools at the impact section: *annular oscillation*, for laminar free jets; *intermittent vortex mechanism,* for jets between laminar and turbulent flow conditions; *turbulent occlusion* for jets presenting an initial turbulence intensity above 2 %; and *droplet entrainment*, for disintegrated jets at the impact section.

About the quantification of air entrained into plunge pools, Oyama et al. (1954) consider that the relationship $\beta_0 = Q_{air}/Q_W$ is a function of Weber, Reynolds and Froude numbers, plus another non-dimensional relationship, given by H_0/d_s, H_0 being the height of the jet freefall and d_s the jet diameter at throwing section.

Baxter (1955) disagrees with the previous authors, affirming that β_0 depends only from one parameter with physical dimensions — $(V_s d_s)^2$, where V_s is the jet mean velocity at the throwing section.

Henderson et al. (1970) consider that β_0 depends from the relationship between diameters of the free jet at throwing section into atmosphere, d_s, and at impact with plunge pool free surface, d_0.

Lin & Donnelly (-), in Ervine et al. (1980), consider that β_0 depends only from the Froude number.

As a possible justification for the above diverging proposals, it is worth mentioning a subsequent study by Ervine & Falvey (1987). These authors consider turbulence intensity as a determining factor in the behaviour of spillway free jets. The fact that none of the previously mentioned authors took into account the considerable influence of this parameter as a possible cause for free jet aeration, may be an explanation for the diverging interpretations for air entrainment advanced in earlier researches.

According to Ervine & Elsawy (1975), the quantity of air entrained into plunge pools by two-dimensional free-falling jets can be estimated by:

$$\beta_0 = 0.13 \left(\frac{H_0}{b_0} \right)^{0.446} \left(1 - \frac{V_{0,min}}{V_0} \right) \tag{1}$$

where β_0 = rate of air to water flow entering the plunge pool; H_0 = height of freefall of the jet; b_0 = jet thickness at entry section; $V_{0,min}$ = minimum mean jet velocity at entry section for air entrainment to occur; V_0 = mean jet velocity at entry section. According to Ervine (1976), with basis on experimentally obtained data, it can be estimated that $V_{0,min} \approx 1.0$ to 1.1 m/s.

In a subsequent study carried out by Ervine & Falvey (1987), another equation was proposed for estimating the quantity of air entrained into pools by free jets, in which a coefficient K was included, to consider the influence of free jet turbulence intensity in the amount of entrained air:

$$\beta_0 = K \sqrt{\frac{s}{d_0}} \tag{2}$$

where K = empirically obtained parameter (Tab. 1); s = jet plunge length from throwing section to pool free surface; d_0 = diameter of circular jets at impact, or jet thickness for rectangular jets. Values of K constant as a function of jet turbulence are presented in Table 1, as proposed by Ervine & Falvey (1987).

Table 1. Constant K values for Equation 2, as proposed by Ervine & Falvey (1987)

Turbulence	Circular jets	Rectangular jets	Application limit
Rough turbulent	0.40	0.20	$s/d_0 \leq 50$
Moderate turbulent	0.30	0.15	$s/d_0 \leq 100$
Smooth turbulent	0.20	0.10	$s/d_0 \leq 200$

Ervine & al. (1997) suggest a modification in Equation 2, in which the factor $(1-V_{0,\min}/V_0)$ from Equation 1 is also included, resulting in:

$$\beta_0 = K\sqrt{\frac{s}{d_0}}\left(1 - \frac{V_{0,\min}}{V_0}\right) \tag{3}$$

Equations 1-3 were obtained for partially developed or undeveloped free jets at plunge pool impact section, meaning that a compact water core still subsists in the central zone of the jet cross section. Research work by Bohrer & Abt (1996) allowed to analyse the influence of jet development (partially and fully developed jets) in the quantity of air entrained into plunge pools, the two following equations having been experimentally deduced:

$$\overline{C}_0 = 0.123 \ln\left(\frac{V_s^2 H_0}{g b_s l_s}\right) + 0.175 \text{, partially developed jets} \tag{4}$$

$$\overline{C}_0 = 0.095 \ln\left(\frac{V_s^2 H_0}{g b_s l_s}\right) + 0.393 \text{, fully developed jets} \tag{5}$$

where $\overline{C}_0$ = initial air concentration at entry section to plunge pool; V_s = mean jet velocity at throwing section; H_0 = height of jet freefall; b_s = jet thickness at throwing section; l_s = jet width at throwing section; and g = gravitational acceleration.

3 SUBMERGED JETS DIFFUSION CONSIDERING AIR ENTRAINMENT

In Figure 1, a schematic representation of the flow pattern associated with the diffusion and floor deflexion of a submerged jet is given, and identified in it are the main variables for this analysis of submerged aerated jets.

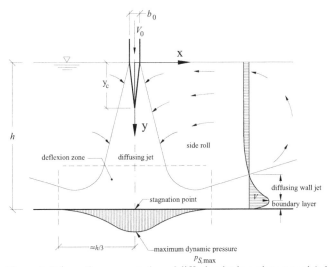

Figure 1. Schematic representation of diffusing jet in a plunge pool (adapted from Kraatz 1989).

127

In free jet spillways, energy dissipation takes place fundamentally in the plunge pool located downstream of the dam, where submerged jet diffuse through the water cushion until its deflection occurs due to the presence of the pool floor.

Concerning the diffusion process of submerged jets formed by either only water or air-water mixtures, Homma (1953) considers that disturbances produced by presence of air bubbles in the latter case lead to more pronounced diffusion angles.

Likewise, Cabelka (1955) considers that air bubbles are entrained downwards due to water flow momentum at impact. When a certain depth y is reached, the effect of buoyancy overcomes momentum, so the bubbles start moving laterally to the jet outer zone, where the flow velocity is smaller. Reaching the shear zone, the air bubbles are caught by existing side rolls (Fig. 1) and invert their movement towards the free surface. Since air bubbles present different velocities from water phase, they produce an increase in turbulence and, consequently, an increase in energy dissipation, resulting in larger diffusion angles for submerged jets in aerated pools than the corresponding diffusion angle in non-aerated pools.

Spurr (1985) and Rahmeyer (1990) also support the explanation advanced by Cabelka (1955).

Kobus (1984) considers that the most significant parameters in the amount of air entrained into plunge pools are: momentum of free jet at impact, which determines the conditions in which air in the free shear layer is more or less pushed into the water cushion; and buoyancy forces, which limit bubble penetration and acts as retarding forces on submerged jets.

Armengou (1991) stressed the differences on the curves associated to mean dynamic pressure distribution on plunge pool floors, when jets with or without air entrainment were considered in experiments. In fact, the inflexion points of non-dimensional gaussian type curves of Figure 2 are further away from the stagnation point in the cases in which air entrainment is considered, revealing greater diffusion angles in these cases.

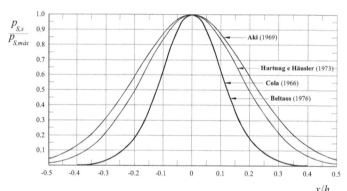

Figure 2. Dynamic pressures on the pool floors for one phase flow (Cola 1966 and Beltaos 1976) and two phase flow (Aki 1969 and Hartung & Häusler 1973), as presented by Armengou (1991).

According to Ervine & Falvey (1987), the presence of air bubbles inside the shear layer limiting the jet diffusion zone leads to a reduction of mean dynamic pressures on pool floors. This situation can be analysed admitting that the two phase flow occurring in the mixing zone can be assimilated to a pseudo-fluid with an apparent density given by $(1-\overline{C}_y)\rho_W$. Considering this hypothesis, the relationship between the mean dynamic pressures at depth y for aerated jets and the corresponding dynamic pressure without air, may be expressed by the following equation:

$$\frac{(p_y)_{with\,air}}{(p_y)_{no\,air}} = \frac{\frac{1}{2}\rho_W\left(1-\overline{C}_y\right)v_y^2}{\frac{1}{2}\rho_W v_y^2} = 1-\overline{C}_y \tag{6}$$

where p_y = maximum mean pressure at depth y; ρ_W = water density; $\overline{C}_y = \beta_y/(1+\beta_y)$ is the mean air concentration at depth y; v_y = maximum jet mean velocity at depth y.

Concerning jet diffusion in a different density medium, Amelung (1995) suggests, based in Kraatz (1965, 89), the following correction to Equation 6 for the specific case of water jets with initial air concentration $\overline{C}_y$, at depth y:

$$\frac{(p_y)_{with\ air}}{(p_y)_{no\ air}} = \left(1 - \overline{C}_y\right)^{1.345} \tag{7}$$

Equation 7 leads to smaller values of the floor pressures relationship than the pseudo-fluid model of Equation 6 advanced by Ervine & Falvey (1987). These authors consider that, for partially developed jets, there still subsists a non-aerated water core at the impact section. As a consequence, in the zone of flow establishment of the submerged jet, the mean dynamic pressure in the core zone is the same as that obtained with non-aerated jets. Immediately downstream of the zone of flow establishment, air concentration is generally quite high (around 40% according to estimations made by Ervine & Falvey 1987) and aerated flow is extended to the whole cross section of the diffusing jet.

As the depth of the penetrating flow increases, a reduction in air concentration is observed, since water flow increases along the diffusing jet path and the volume of air diminishes progressively as air bubbles in the shear layer are transferred to side rolls. Air is then brought to the surface and escapes into the atmosphere.

Amelung (1995) considers that the relationship between air and water flow rates at depth y, represented by β_y, can be estimated for the particular case of plane jets, in a analogous manner to that proposed by Ervine & Falvey (1987) suggested for circular jets.

Concerning the increase of water discharge Q_W with increasing depth y, Hartung & Häusler (1973) present the following equation for plane submerged jets:

$$\frac{Q_{W,y}}{Q_{W,0}} = \sqrt{\frac{2y}{y_c}} = \sqrt{\frac{2y}{C_d b_0}} > 1 \tag{8}$$

where Q_w = water discharge rate; y_c = submerged jet core length; and C_d = diffusion coefficient as defined by Albertson et al. (1948) for semi-infinite pools, or by Cola (1966) for confined pools.

For the diffusion coefficient C_d in Equation 8, the value associated with non-aerated water jets diffusing in confined pools is recommended, which is about 7.2 to 7.3 according to Cola (1966), Beltaos (1976), Aksoy (1975), Kraatz (1989) or Melo (2001).

The air flow, Q_{air}, diminishes with the increase in depth y up to a determined penetration depth, y_p. Downstream of that depth, air will be absent from flow. Ervine & Falvey (1987) admit that this reduction follows a linear trend and suggest the following equation to express the variation of air flow with depth y:

$$\frac{Q_{air,y}}{Q_{air,0}} = 1 - \frac{y}{y_p} \tag{9}$$

where y_p = air bubble penetration depth. For an approximate evaluation of the penetration depth, Ervine & Falvey (1987) advance the following expression:

$$y_p \approx 1.5 \frac{V_0}{v_{air}} b_0 \tag{10}$$

where v_{air} = bubble rise velocity.

In Equation 10 the penetration depth of the air bubbles depends on the mean free jet velocity at impact section, V_0, jet thickness, b_0, and bubble rising velocity, v_{air}. This last parameter may be estimated, as advanced by Ervine & Falvey (1987), as being approximately constant and close to the value v_{air}=0.25 m/s. Admitting a deterministic value to v_{air} is a simplification of the phenomenon, since some significant aspects are being excluded, such as dependency of the rising velocity from bubble geometry and compressibility effects of the air due to pressure variations in plunge pool. This effect is quite significant in prototypes where the pressure variation along the diffusing jet path is considerable.

Associating Equations 8 and 9, it is possible to estimate the value β_y at sections of the diffusing jet downstream of the zone of flow establishment ($y>y_c$) with the following expression:

$$\beta_y = \frac{Q_{air,y}}{Q_{W,y}} = \beta_0 \sqrt{\frac{C_d}{2}} \left(\sqrt{\frac{b_0}{y}} - \sqrt{\frac{y}{b_0}} \frac{v_{air}}{1.5\,V_0} \right) \qquad (11)$$

Considering Equations 7 and 11, an expression can be obtained to estimate the relationship between dynamic pressures on pool floors ($y=h$) with and without air, as a dependent function of jet characteristics at impact - V_0, b_0 e β_0, and diffusion characteristics of submerged jets in confined pools, C_d:

$$C_b = \frac{\left(p_{S,\max}\right)_{with\ air}}{\left(p_{S,\max}\right)_{no\ air}} = \frac{1}{\left(1 + \beta_0 \sqrt{\dfrac{C_d}{2}\dfrac{h}{b_0}} \left(\dfrac{b_0}{h} - \dfrac{v_{ar}}{1.5V_0}\right)\right)^{1.345}} \qquad (12)$$

The pressure relationship presented in Equation 12 will be represented in the following text as C_b, which will be termed as "buoyancy coefficient".

For values of $1,5V_0/v_{ar}$ higher than h/b_0, the buoyancy coefficient C_b given by Equation 12 becomes greater than one, meaning that this equation is valid only when pool depth, h, is higher than air bubbles penetration depth, y_p.

4 POOL DEPTH, JET THICKNESS AND VELOCITY EFFECT ON FLOOR PRESSURES

Equation 12 was applied for a specific range of values of h/b_0 and V_0, to analyse the influence of each of these parameters in the buoyancy coefficient, C_b.

The variation of the diffusion coefficient, C_d, is small for the usually considered range of h/b_0 values for practical purposes in spillways ($10 < h/b_0 < 20$), so it was assumed the constant value $C_d = 7,3$. The variation of the mean air concentration in this analysis goes from 0 to 50 %.

The parameter v_{ar}/V_0 reflects the scale effects associated with the physical modelling of energy dissipation of jets with air entrainment. For the analysis of its influence in C_b, a variation of V_0 between 3 and 30 m/s was admitted. The obtained curves of $C_b = f(\overline{C}_0, V_0)$ are depicted in Figure 3.

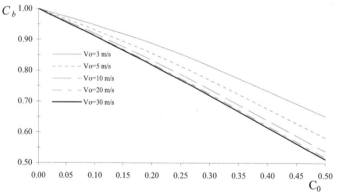

Figure 3. Variation of the buoyancy coefficient C_b as a function of $\overline{C}_0$ and V_0 using Equation 12. (v_{air}=0.25 m/s and h/b_0=10).

The graph of Figure 3 allows verification of the fact that the beneficial influence of air depends on V_0, being more relevant in situations that are typical of prototypes (V_0 values of about 20 to 30 m/s) than in those common in models (V_0 values between 3 and 10 m/s). Moreover, for values of V_0 above 20 m/s, C_b seams to become almost independent from variations in V_0.

For the analysis of the influence of relative thickness between the jet and the pool, h/b_0 was varied from 8 to 20. This range covers most of the practical cases of plunge pool energy dissipators in dam spillways.

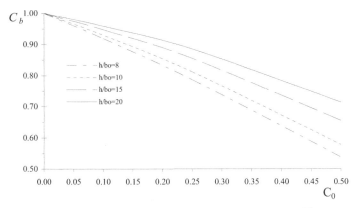

Figure 4. Variation of the buoyancy coefficient C_b as a function of $\overline{C}_0$ and h/b_0 using Equation 12 (v_{air}=0.25 m/s and V_0=10 m/s).

The curves presented in Figure 4 reveal a certain dependency of the buoyancy coefficient, C_b on the relative thickness, h/b_0, presenting C_b a progressive reduction with growing values of h/b_0. This situation reflects the reduction of the air concentration due to the joint effect of water flow increase and the volume of air reduction with depth increase.

According to this analysis and for the range of values considered, the coefficient C_b decreases from 1.0 to about 0.5 as air concentration increases from 0 to 50 %.

5 EXPERIMENTAL STUDIES

5.1 *Preliminary considerations on scale effects and experimental set-up*

As pointed out by Ervine et al. (1997), it is difficult to model free-falling jets, because surface tension and turbulence forces considerably influence break-up and air entrainment.

Considering that the purpose of the study was the analysis of the pressure field on the plunge pool floors for different situations of air concentration at impact, it was decided to analyse the problem using a submerged jet exiting from a nozzle directly into the pool, instead of using a free jet. A Photo with a general view is included in Figure 5 and a scheme of the experimental facility is presented in Figure 6.

Figure 5. General view of the experimental installation.

131

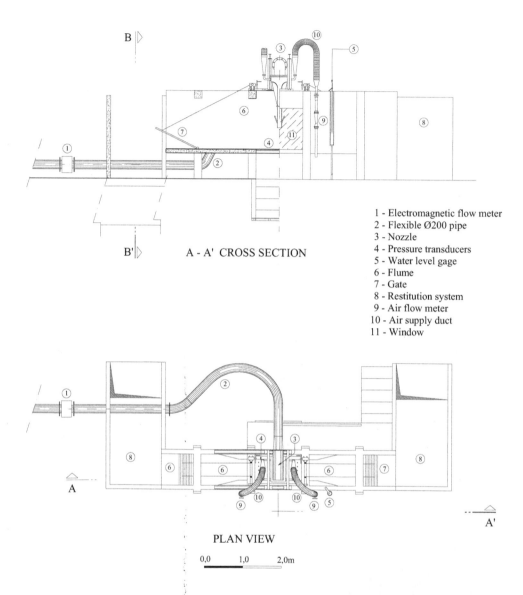

B ▷

A - A' CROSS SECTION

1 - Electromagnetic flow meter
2 - Flexible Ø200 pipe
3 - Nozzle
4 - Pressure transducers
5 - Water level gage
6 - Flume
7 - Gate
8 - Restitution system
9 - Air flow meter
10 - Air supply duct
11 - Window

B' ▷

PLAN VIEW

0,0   1,0   2,0m

Figure 6. Main features of the experimental installation.

The air was introduced into the water flow through two slots inserted in the nozzle in a depressed zone of the water flow, as depicted in the scheme of Figure 7. This means that the flow inside the nozzle is already composed of an air-water mixture upstream from the entrance in the pool, as shown in the photograph of Figure 8.

132

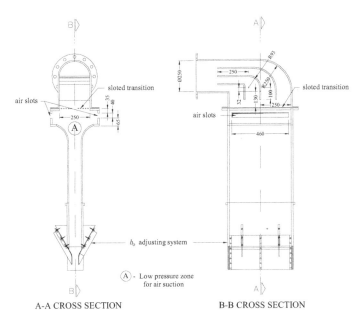

A-A CROSS SECTION   B-B CROSS SECTION

Figure 7. Schematic representation of the jet nozzle and aeration slots.

Figure 8. General aspect of the flow formed by the submerged air-water jet.

The submerged jet air concentration at the entrance of the pool was imposed through an independent control and measurement of the water and air flows (Figs 6-7).

Another important aspect that may lead to scale effects in experimental air-water flow studies is the difficulty in respecting the similarity relationships of bubble's size. In the described study this was not controlled.

5.2 *Experimental results concerning the pressure field*

Experimental tests were carried out in which an evaluation of the entrained air effect in the pressure field was made by comparing the maximum pressures on the floor produced by aerated jets with those produced by non-aerated jets.

The following comments on the obtained results are presented:
– the experimental curve expressing the variation of buoyancy coefficient, C_b, with air concentration is presented in Figure 9, indicating that, for $\overline{C}_0=30$ %, the mean dynamic pressure under the jet axis is 76 % of that observed for an equivalent jet with $\overline{C}_0=0$ %;

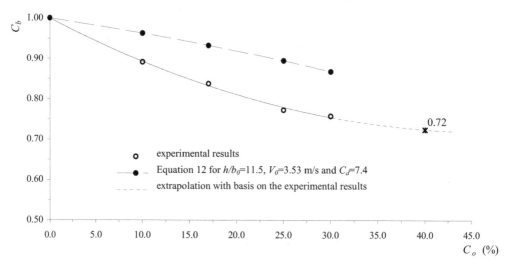

Figure 9. Variation of the buoyancy coefficient, C_b, with the mean air concentration at the pool entrance, $\overline{C}_0$, for the particular case of $h/b_0=11.5$.

– the effect of air concentration variation in dynamic pressures decreases for increasing values of $\overline{C}_0$;
– despite the limited range considered in the experimental tests for the air concentration (0 to 30 %), the quantification of $\overline{C}_0$ effect on the plunge pool floor revealed a significant attenuation of dynamic mean pressures.
– application of Equation 12 using the parameter values of the considered experimental conditions, leads to conservative results in terms of dynamic pressure on the pool floor, meaning it underestimates the energy dissipation of the diffusing submerged air-water jet;
– due to limitations of the experimental facility it was not possible to carry out measurements using jets with air concentrations above 30 %, therefore it was not possible to obtain experimental results for jets with air concentrations of about 40 %, which is a value advanced by Ervine & Falvey (1987) as characteristic in turbulent free jets in prototype spillways;
– an extrapolation of the experimental results was made by regression analysis to estimate the value of C_b for $\overline{C}_0=40$ %, and $C_b\approx0.72$ was obtained.

6 CONCLUSIONS

The following conclusions are outlined from the literature analysis relating free jet air entrainment:
– internal turbulence of free jet flows is a decisive parameter in the development degree of jets at impact, lateral spreading along the trajectory in atmosphere and break-up length;

134

- air entrainment by turbulent free jets results mostly from turbulent occlusion associated to jet rough surface and, in a smaller degree, from the air boundary layer developed around the water jet;
- air is entrained into plunge pools as an air-water mixture, Equations 1-5 allowing an estimation of the ratio air to water flow, given by $\beta_0 = Q_{air}/Q_w$, for undeveloped, partially developed or fully developed turbulent free jets.

Regarding submerged jet diffusion with air entrainment, the following conclusions are presented:
- presence of air in plunge pools allows a significant reduction of the mean dynamic pressures in the pool floor;
- based on a literature analysis, Equation 12 is proposed for estimation of the effect of air entrained into pools on the dynamic mean pressures;
- the effect of air entrained into plunge pools by plunging jets was quantified experimentally up to a concentration of 30 % at impact section, and a dynamic pressure reduction on the floor of 24% was obtained;
- with due reserves, an extrapolation of the experimental results was carried out for an air concentration at impact of 40 %, the corresponding reduction of the mean dynamic pressure in the pool floor being estimated in 28 %.

As a final remark, it is considered that additional experimental studies are required to analyse eventual influence of other parameters, such as pool depth, h, or the velocity at impact, V_0, on the buoyancy coefficient, C_b.

Though it would be highly desirable to perform studies for validation of the presented results based on prototype measurements, the associated costs are a difficult obstacle to overcome, so further research will be limited to laboratory experimental studies.

ACKNOWLEDGMENTS

The author would like to thank Mr. Matias Ramos from the National Laboratory of Civil Engineering, Lisbon, Portugal, and Prof. António Pinheiro from the Technical University of Lisbon, Portugal, for all the help and guidance conceded during the course of this study. The author also wishes to thank Dr.-Ing. Martin Amelung the useful exchange of ideas and information, which proved essential for this work. Finally, a word of recognition to the Portuguese Foundation for Science and Technology (FCT) for the financial support conceded through the Praxis XXI Programme.

REFERENCES

Aki Schuichi 1969. "Jiyu rakka suimuyaku no mizu-kusshon koka mi kansuru kenkyu" (Study of the efficiency of water cushioning in free-falling jets), *Denryoku chuo kenkyujo hokoku* (Journal of the Research Institute for Hydroelectric Power Stations).

Aksoy, S. 1975. "Örtliche Felskolke unterhalb der Hochwasserentlastung hoher Talsperren". *Wissenschaftlich-Technische Forschungsanstalt der Türkei*, Ankara (original in Turkish).

Albertson, M.L.; Dai, Y.B.; Jensen, R.A. & Rouse, H. 1948 - "Diffusion of submerged jets". *Proceedings ASCE*, 74, December, 1571-1596.

Amelung, M. 1995. *Auskolkung Klüftiger Felssohlen durch Entlastungsstrahlen*. Ph.D. Thesis, Technischen Universität Carolo-Wilhelmina, Shaker Verlag, Braunschweig, Germany, April.

Armengou, J. 1991. *Vertido Libre por Coronación en Presas Bóveda. Análisis del Campo de Presiones en el Cuenco Amortiguador*. Tesis Doctoral, Universität Politècnica de Catalunya, Junio.

Beltaos, S. 1976. "Oblique impingement of plane turbulent jets". *Journal of the Hydraulics Division, ASCE*, Vol. 100, No. HY10, September, 1177-1192.

Bohrer, J.G. & Abt, S.R. 1996. *Plunge Pool Velocity Decay of Rectangular Free Falling Jets. Dam Foundation Erosion. Phase II . Clear Water Experiments, 1:3 Scale Model Facility*. Bureau of Reclamation, Denver.

Bollaert, E. & Schleiss, A. 2001. "Air bubble effects on transient water pressures in rock fissures due to high velocitry jet impact". 29th IAHR Congress, Beijing, Vol. II, 538-543.

Cabelka, J. 1955. "The losses of mechanical energy of the overfall jet on the spillway sections of dam". *VI IAHR Congress*, Den Haag, Vol. 3, C27-1 - C27-10.

Cola, R. 1966. "Diffusione di un getto piano verticale in un bacino d'acqua d'altezza limitata". *L'Energia Elettrica*, N°11, 649-667.

Ervine, D.A. 1976. "The entrainment of air in water". *Water Power & Dam Construction*, December, 27-30.

Ervine, D.A. & Elsawy, E.M. 1975. "Model scale effects in air-regulated siphon spillways". *BHRA Symposium on Siphons and Siphon Spillways*, London, May, Paper B2.

Ervine, D.A. & Falvey, H.T. 1987. "Behaviour of turbulent water jets in the atmosphere and in plunge pools". *Proc. Inst. Civ. Engrs.*, Part 2, March, 295-314.

Ervine, D.A.; Falvey, H.T. & Withers, W. 1997. "Pressure fluctuations on plunge pool floors". *Journal of Hydraulic Research*, Vol. 35, No.2, 157-279.

Ervine, D.A.; McKeogh, E. & Elsawy, E.M. 1980. "Effect of turbulence intensity on the rate of air entrainment by plunging jets". *Proc. Inst. Civ. Engrs.*, 69, June, 425-445.

Falvey, H.T. 1980. *Air-water Flow in Hydraulic Structures.* U. S. Department of the Interior, Water and Power Resources Service, Engineering Monograph n°41, Denver.

Hartung, F. & Häusler, E. 1973. "Scours, stilling basins and downstream protection under free overfall jets and dams". *11th ICOLD Congress*, Madrid, Vol. II, Q.41, R.3.

Henderson, J.B.; McCarthy, M.J. & Molloy, N.A. 1970. *Proc. Chemeca Conf.*, Australia, Section 2, p.86.

Homma, M. 1953. "An experimental study on water fall". Proceedings Minnesota International Hydraulics Convention, IAHR, Minneapolis, 477-481.

Kobus, H. 1984. "Local air entrainment and detrainment". *Symposium on Scale Effects in Modelling Hydraulic Structures*, Technische Akademie Esslingen, paper 4.10.

Kraatz, W. 1965. "Flow characteristics of a free circular water jet". *XI AIRH Congress*, Volume I, Leningrad, paper 1.44.

Kraatz, W. 1989. "Flüssigkeitstrahlen". In Bolrich et al, *Technische Hydromechanik 2*, VEB Verlag für Bauwesen, Berlin.

Lin, T.J. & Donnelly, H.G. (-). "Gas bubble entrainment by plunging laminar liquid jets". *A.I.Ch.E.J.*, 12, No.3.

McKeogh, E.J. 1978. *A Study of Air Entrainment using Plunging Water Jets.* Ph.D Thesis, Queen's University, Belfast, October.

Melo, J.F. 2001. *Acções Hidrodinâmicas em Soleiras de Bacias de Dissipação de Energia por Jactos.* Ph. D. Thesis. Instituto Superior Técnico, Technical University of Lisbon.

Oyama Y.; Takashima Y. & Idemura H. 1954. "Air entrainment phenomena by jets". *Repts. Sci. Research Inst.,* Japan, 29, p.344

Rao, N.L. & Kobus, H.E. 1975. "Characteristics of self-aerated surface flows". *Water and Waste Water, Current Research and Practice.* Vol. 10, Erich Schmidt Verlag, Berlin.

Rahmeyer, W. 1990. "The effect of aeration on scour". *Hydraulic Engineering*, Chang, H.H. & Hill J.C., 1990, 531-536.

Spurr, K.J.W. 1985. "Energy approach to estimating scour downstream of large dams". *Int. Water Power and Dam Construction*, 37, July, 81-89.

The influence of plunge pool air entrainment on the presence of free air in rock joints

E. Bollaert
Laboratory of Hydraulic Constructions, Swiss Federal Institute of Technology, Lausanne, Switzerland

ABSTRACT: High-velocity plunging water jets, appearing at the downstream end of dam weirs and spillways, can create scour of the rock. The prediction of this scour is necessary to ensure the safety of the toe of the dam as well as the stability of its abutments. A physically based engineering model has been developed at the Laboratory of Hydraulic Constructions for evaluation of the ultimate scour depth. This model is based on experimental measurements of water pressures at plunge pool bottoms and inside underlying rock joints. The pressures inside the joints revealed to be of highly transient nature and governed by the presence of free air. The amount of free air is a function of the instantaneous pressure inside the joint and the instantaneous air content in the plunge pool. This statement is based on an experimental and numerical determination of the governing "wave celerity-pressure" relationships. These relationships are defined by the ideal gas law and Henry's law.

1 INTRODUCTION

High-velocity plunging water jets, appearing at the downstream end of dam weirs and spillways, can create scour of the rock. Scour is often predicted by empirical or semi-empirical formulae, developed from physical models or prototype observations. These formulae are not fully representative because they cannot describe all of the physical effects involved. Above all, the characteristics of pressure wave propagation in the fissures of the jointed rock mass are unknown. Therefore, an experimental facility has been built at the Laboratory of Hydraulic Constructions that measures pressure fluctuations at plunge pool bottoms and simultaneously inside underlying rock joints. The facility simulates the falling jet, the plunge pool and the fissured rock mass. Prototype jet velocities are impacting through a water cushion onto a plunge pool bottom, generating so pressure fluctuations that can enter the underlying rock joint. The water pressures are measured simultaneously at the pool bottom and inside the simulated rock joint. As such, a direct relationship between pressures in plunge pools and pressures that are transferred inside a jointed rock mass has been outlined. A more detailed description of the facility can be found in Bollaert & Schleiss (2001) and Bollaert (2002).

The water pressures have been measured inside two basic one-dimensional rock joint configurations (Figure 1) and are governed by the propagation of pressure waves. The wave celerity has been determined based on peak frequencies in the power spectral density and based on direct pressure pulse propagation speed measurements. The wave celerity and, thus, free air content is found to depend on the instantaneous pressure value inside and can be expressed by means of a "wave celerity-pressure" relationship. It appears that the general form of this relationship can be described by two physical laws: the *ideal gas law* and *Henry's law*. Moreover, the exact position of the relationship continuously changes with time as a function of the air content in the plunge pool. A comparison is made of "wave celerity-pressure" relationships for both submerged and impinging jets and for two differently shaped rock joints.

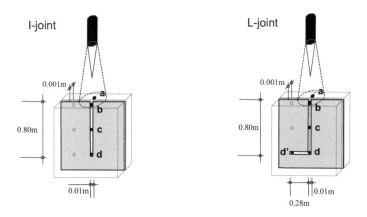

Figure 1. Sketch of the tested one-dimensional rock joint configurations, corresponding to the I-joint and the L-joint. Sensor (a) measures the pressures at the plunge pool bottom, while the sensors (b), (c), (d) and (d') measure pressures inside the rock joint.

2 TYPES OF JET IMPACT

Dynamic pressures can be generated by core jet impact, appearing for small plunge pool depths Y, or indirectly by turbulent shear layer flow, appearing for ratios of pool depth Y to jet thickness D_j higher than 4 to 6 (for impinging jets). The resulting pressure fluctuations at the water-rock interface are completely different. Terminology therefore distinguishes between *core jet impact* (Figure 2a) and *developed jet impact* (Figure 2b).

As can be seen in Figure 2a, core jet impact combines a small and not fully developed shear layer with direct jet impact under the jet's centreline. The corresponding pressures have a high mean value with small fluctuations around this mean value (low RMS = root-mean-square value). The developing turbulent eddies are small and result in the appearance of high frequencies in the power spectral content of the pressure signal. On the other hand, developed jet impact has no core anymore and generates a large shear layer. The mean dynamic pressure decreases, while the pressure fluctuations increase. The power spectral content of the pressure fluctuations concentrates its energy towards the intermediate-scaled turbulent eddies.

In other words, the excitation capacities of the jet are completely different for core respectively developed jet impact. This has a great influence on the pressure fluctuations in the underlying rock joint.

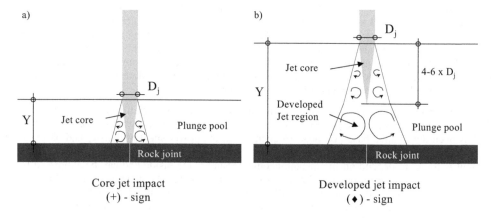

Figure 2. Types of jet diffusion and impact at the pool bottom: a) core jet impact (+ symbol) and b) developed jet impact (♦ symbol) (Bollaert & Schleiss, 2001).

3 WAVE CELERITY DETERMINATION

The measured water pressures have been recorded by a series of pressure sensors: sensor (a) is located at the plunge pool bottom, close to the rock joint entrance and quasi under the jet's centerline, while the sensors (b), (c), (d) and (d') are located inside the closed-end rock joint (Figure 1). Pressure measurements have been made simultaneously at these locations, in order to establish a correlation.

The highly transient two-phase behavior of the pressures inside the rock joints depends on the amount of free air inside the joints. Because free air has a significant impact on the resonance characteristics of a transient system, the amount of free air can be evaluated in an indirect manner based on the fundamental resonance frequency of the system in question. Determination of the peak frequency in the power spectral content or in the transfer function of the pressure signals defines a fundamental resonance frequency f_{mean}, which corresponds to a fundamental or *mean wave celerity* c_{mean}. This is obtained by assuming a perfectly open-closed boundary ($\lambda/4$) resonator system, for which the following relation holds:

$$f_{mean} = (1 + 2n) \cdot \frac{c_{mean}}{4L} \tag{1}$$

L is the total length of the closed-end joint. For n = 0, the first or fundamental harmonic is obtained. In the following, it is assumed that the lowest peaks in the power spectral content correspond to the fundamental resonance frequency, i.e. n = 0. By knowing the length L of the joint, this defines the mean wave celerity c_{mean}. Hence, the relationship between the wave celerity and the amount of free air of a gas-liquid mixture results in the mean free air content inside the joint α_{mean}. The governing equation is presented in the next paragraph.

The celerity of a pressure wave, however, is not a constant but depends on the absolute value of the governing pressure. This is due to a change in free air concentration in the joint as a function of pressure, which is governed by the ideal gas law and Henry's law. The result is that the peak frequencies in the power spectral content are smeared out over a whole range of frequencies. Moreover, the pressure is not constant throughout the length of the joint and continuously changes as a function of time. Hence, only a space and time averaged, very rough estimate of free air concentration can be accounted for with the method of peak spectral frequencies.

However, the wave celerity can also be defined by making a correlation between different pressure sensors. This determines the *instantaneous wave celerity* between the sensors in question as a function of the governing pressure that holds at that moment between these sensors. A plausible time-averaged correlation between two sensors could not be defined, again due to the continuously changing amount of free air inside the rock joints. Therefore, it was decided to establish sensor-to-sensor correlations only for distinct pressure pulses entering the rock joints, and not for a certain time period.

These correlations are based on direct measurements of the propagation speed of the pressure pulse that travels through the joint. Such a pulse enters the rock joint and is measured successively at the sensors (b), (c) and (d). The distance Δl and the time difference Δt between two sensors define the travel speed of the wave $c = \Delta l/\Delta t$. The pressure values during this time interval define an average pressure value p_m.

This procedure has been systematically performed for a series of test runs. For every test run, different pressure levels have been analysed. Each pressure level was checked several times on its corresponding propagation speed. It has to be noted that the degree of precision of this process increases with decreasing time step of pressure measurements, i.e. with increasing pressure acquisition rate. As the importance of free air was not evident at the start of the test runs, most of them have been established at an acquisition rate of only 1'000 Hz. It revealed that, at high pressures, this rate was not sufficient to accurately predict the wave celerity-pressure relationship. Fortunately, some test runs were performed at rates of 5'000 to 20'000 Hz. These runs yielded sufficiently accurate results, even at high pressures.

An example of the followed procedure is presented in Figure 3 for the I-joint. The plunge pool depth Y is 0.40 m and the jet velocity at impact V_j is 20 m/s. As can be seen, a pressure pulse enters the rock joint. This is followed by a pressure rise at the middle of the joint (sensor (c)) and finally at the end of the joint (sensor (d)). The time difference Δt between the pressure rise at the middle and at the end can be estimated within a precision of +/- 0.00005 sec. During this time interval, the pressure at the middle of the joint progressively increases. By defining the pressure values at the beginning and at the end of the time interval, a mean pressure value p_m can be defined. This pressure value is considered as characteristic for the pressure pulse that has travelled from one sensor to the other. It can be defined within +/- 0.5 m of pressure head. Hence, the wave celerity of the pulse c_{pulse} is determined as follows:

$$c_{pulse} = \frac{\Delta(sensor(c) - sensor(d))}{\Delta t} = \frac{0.375m}{\Delta t} \qquad (2)$$

For the present example, the time interval Δt is estimated at 0.0025 seconds. With a Δl of 0.375 m, this results in a pulse propagation speed of 150 m/s. The corresponding mean absolute pressure head p_m equals 17 m. The obtained precision of the wave celerity calculation is +/- 6 m/s. This process has been repeated at different pressure levels.

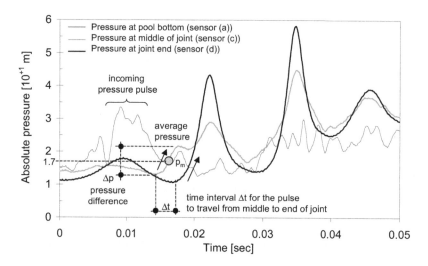

Figure 3. Determination of the instantaneous wave celerity c_{pulse} of a pressure pulse that enters the I-shaped rock joint. The pool depth is 0.40 m and the jet velocity is 20 m/s.

4 THEORY OF AIR CONTENTS IN PLUNGE POOLS AND ROCK JOINTS

No measurements have been made of the air content at the point of jet impact in the plunge pool. Hence, evaluation of this parameter has been performed based on theoretical expressions that are particularly valid for circular impinging jets. Furthermore, the air content at the point of impact of the jet has been used as value at the rock joint entrance. Although real plunge pools may exhibit a significant diffusion of the air content through the water depth, the very small water depths used in the experimental facility may justify the here-adopted approach.

Secondly, the free air content of the water inside the rock joints can be derived from the mean or instantaneous wave celerity. The wave celerity is greatly reduced when a small amount of air is present in the form of free bubbles (Wylie & Streeter, 1978). As such, a small free air content in a liquid can produce wave speeds that are less than the speed of sound in the air alone. The free air bubbles could be imagined as springs that are compressed by a pressure pulse. The spring accelerates the liquid, which, in turn, compresses the next spring. A pressure pulse can so be transmitted at a lower velocity than in a homogenous liquid, in which the transmission is done from one molecule to the next. Two basic physical laws govern the relation between the amount of free air and the wave celerity: the ideal gas law and Henry's law.

4.1 Plunge pool air content

The air content at the point of jet impact in the plunge pool is dominated by the turbulence intensity Tu of the issuing jet (Falvey & Ervine, 1988). The magnitude of surface disturbances is related to the square root of the plunge length L rather than to gravity considerations. Such a relationship has been established by Ervine (1976) and Bin (1984) for circular plunging jets:

$$\beta = (0.2 \text{ to } 0.4) \cdot \sqrt{\frac{L}{D_j}} \cdot \left(1 - \frac{1}{V_j}\right) \qquad \text{Ervine (1976)} \qquad (3)$$

$$\beta = 0.05 \cdot Fr_j^{0.56} \cdot \left(\frac{L}{D_j}\right)^{0.4} \qquad \text{Bin (1984)} \qquad (4)$$

Bin (1984) also performed a correlation of the data of Van de Sande & Smith (1973) resulting in the following equation for circular plunging jets:

$$q_a = 0.015 \cdot \left(\frac{D_j^2 \cdot V_j^3 \cdot \sqrt{L}}{\sin \theta}\right)^{0.75} \qquad \text{Van de Sande & Smith (1973)} \qquad (5)$$

The turbulence intensity Tu of the issuing jets is indirectly incorporated in these expressions by means of the magnitude of the proportionality coefficient. It seems convenient to define a best-fit curve based on the above expressions. This has been done for different L/D_j values, and corresponds to air contents of 15 % to 35 % at jet velocities of 10 m/s, and air contents of 40 % to 60 % at jet velocities of 30 m/s. This means that, at near-prototype jet velocities, the physical maximum air content is approached, even under laboratory conditions.

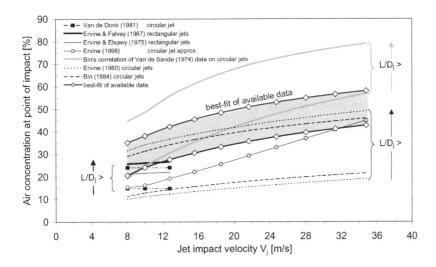

Figure 4.   Summary of expressions for the air concentration at point of jet impact in a plunge pool. The air content is given as a function of the jet velocity at impact V_j and of the ratio of the fall length over jet diameter L/D_j, for both circular and rectangular plunging jets (Bollaert, 2002).

4.2 The ideal gas law

The wave celerity in a certain mixture can be expressed as a function of its bulk modulus and its density as follows:

$$c = \sqrt{\frac{K_{mix}}{\rho_{mix}}} \qquad (6)$$

By considering a control volume $\forall$, consisting of a liquid volume $\forall_{liq}$ and an air volume $\forall_{air}$, the bulk modulus of elasticity of the mixture K_{mix} can be written as a function of the bulk modulus of elasticity of the individual components, K_{liq} and K_{air}:

$$K_{mix} = \frac{K_{liq}}{1 + (\forall_{air}/\forall) \cdot (K_{liq}/K_{air} - 1)} \qquad (7)$$

In general, there is a strong influence of interfacial heat, mass and momentum transfer between the two phases. For small air contents, such as in a moderately bubbly liquid, the density of the mixture is essentially that of the liquid, but the compressibility is strongly dominated by the air. An increase in pressure results in an increase in wave celerity, with the value of the liquid as upper bound. The relationship between wave celerity and pressure is expressed by the *ideal gas law*. This law relates the pressure p, the volume $\forall_{air}$ and the absolute temperature T of a fixed mass of gas as follows:

$$p \cdot \forall_{air} = n \cdot R \cdot T \qquad (8)$$

in which R is the universal gas constant and n is the amount of molecules (= number of moles). For a constant absolute temperature T and a constant mass of air, an increase in pressure conducts to a decrease in volume:

$$\forall \propto \frac{1}{p} \qquad (9)$$

The ideal gas law is approximately true for all gases under normal laboratory conditions (room temperature and standard atmospheric pressure) and all gases obey this law more and more closely as the pressure becomes closer to the atmospheric pressure. The volume of the mass of free air that is available in a liquid is inversely proportional to the pressure of the liquid. By accounting for the elastic properties of the wall boundaries, the following expression is obtained for the wave celerity c_{mix} [m/s] of a homogeneous mixture of free air and liquid, in which D [m] is the hydraulic diameter, E [N/m^2] is the Young's modulus of elasticity of the boundary, e [m] is the width of the boundary, R [-] is the universal gas constant, T [K] is the absolute temperature, p [N/m^2] is the absolute pressure and m [kg/m^3] is the mass of free air per unit of volume of the mixture:

$$c_{mix} = \sqrt{\frac{\dfrac{K_{liq}}{\rho}}{1 + \dfrac{K_{liq} \cdot D}{E \cdot e} + \left(\dfrac{mRT}{p}\right) \cdot \left[\left(\dfrac{K_{liq}}{p}\right) - 1\right]}} \qquad (10)$$

The term K_{liq}/p is very large compared to the unity and the latter may be dropped. It has to be noted that the ideal gas law only describes the volumetric change with pressure of a constant mass of free air, i.e. during pressure changes no additional mass of free air is generated.

The wave celerity as defined in equation (10) can be visualized as a function of the absolute pressure head for different free air contents and for different boundary characteristics. In case of rock joints, the hydraulic diameter D can be expressed as twice the joint width, and the thickness of the boundary e is considered as infinite. Therefore, the term expressing the elasticity of the boundary in equation (10) can be dropped. Figure 5a presents the wave celerity of a homogeneous air-water mixture as a function of the absolute pressure, for different air contents.

142

The indicated air volumes correspond to standard atmospheric pressure conditions and room temperature. Each curve is valid for a constant mass of free air, only the volume of the mass of free air changes with pressure.

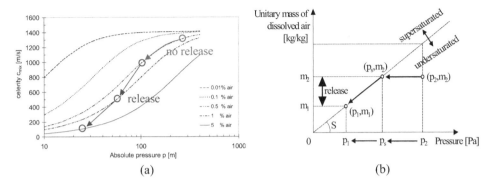

(a)   (b)

Figure 5.   a) The wave celerity of an air-water mixture as a function of its absolute pressure head. The indicated air volumes are valid at atmospheric pressure.; b) Henry's law for air release and re-solution. A sudden pressure drop from pressure p_2 to pressure p_1 releases a unitary mass of air defined by $m_2 - m_1$.

4.3 Henry's law

Beside the presence of free air, most liquids also contain some air in solution. Although the dissolved air doesn't influence the compressibility of the mixture, it represents a potential free air volume whenever air release can happen. Air release is highly depending on pressure, and every system subjected to severe pressure changes is sensible to it. Therefore, the dissolved air content should be quantified, as well as its potential to be released.

The mass of dissolved gas in a certain volume of liquid was determined in 1801 by the English chemist William Henry and is called *Henry's law*. This law states that the solubility of a gas in a liquid is proportional to the partial pressure of the gas above the liquid, provided that the temperature and molecular structure of the mixture remain constant. Henry's law relates the maximum concentration of dissolved air to the air saturation pressure by means of the Bunsen solubility constant S:

$$\frac{\forall'_{air}}{\forall} = m = S \cdot \frac{p_{s,air}}{p_0} \tag{11}$$

in which $\forall$ stands for the total volume of the mixture and the volume of dissolved air $\forall'_{air}$ is reduced to standard pressure conditions (= 1 atm) and a temperature of 25° C. The unitary mass of dissolved air m [kg/kg] is defined as the product of the Bunsen solubility constant S with the ratio of the air saturation pressure $p_{s,air}$ to the standard atmospheric pressure p_0 (= 1 atm = 1 bar $\cong 10^5$ Pa). For air that is dissolved in water, S equals 0.0184. The pressures are expressed in absolute values.

The mass of dissolved air that can be released from the liquid is then defined as follows (Figure 5b). Assume that the air-water mixture is in an undersaturated equilibrium at a pressure p_2 and a unitary mass of dissolved air m_2. A sudden decrease in pressure from p_2 to p_1 involves air release, provided that p_1 is smaller than the saturation pressure p_s of the mixture. The unitary mass of air that is released by the liquid to reinstall the dynamic equilibrium is given by (Schweitzer & Szebehely, 1950):

$$m_2 - m_1 = S \cdot \frac{(p_s - p_1)}{p_0} \tag{12}$$

143

Henry's law is a dynamic equilibrium relation, which is able to response in both forward and reverse sense: a pressure increase results in a higher air content, while a pressure decrease conducts to a lower air content. An application is the production of soft drinks and champagne. In both cases, carbon dioxide is dissolved in the liquid under pressure. When the bottle is opened, the pressure above the solution is released, the solubility of the gas is suddenly reduced, and the gas comes out of solution. A hydraulic application can be found in systems where significant free air is entrained in the flow environment at relatively low pressures. This can happen for example at a dropshaft or a siphon spillway. A downstream rise in pressure, due to for example a closed conduit, results in a supersaturated liquid.

Gas release needs a certain time period, called the incubation period. For the release of air from water, the incubation period to obtain equilibrium is typically several seconds (Schweitzer & Szebehely, 1950). Moreover, the rate of evolution of the air is directly depended upon a number of parameters, such as the degree of agitation of the liquid, the void fraction of the free gas, the importance of the pressure drop (degree of supersaturation), the solubility constant S, the boundary conditions, and the size and the location of nuclei. The proportionality between the rate of evolution and the degree of supersaturation involves that it is an exponential function of time. The precise expression and rate of air release in a certain flow environment is a very complex and dynamic problem. For every particular geometric and hydrodynamic situation, a different law applies.

As a conclusion, Henry's law describes the change of mass of free air as a function of pressure conditions and, thus, the corresponding "wave celerity – pressure" relationships jump from on e curve of constant mass of free air to another (see Figure 5a). The here described wave celerity measurements indicate a quasi-instantaneous air release and re-solution as a function of pressure changes inside the joints. This is in contradiction with the incubation periods that are normally encountered in hydraulic systems and is probably due to the particular geometric and hydraulic conditions that govern in a narrow rock joint.

5 INFLUENCE OF POOL AERATION ON MEAN AIR CONCENTRATION IN JOINTS

5.1 The I-joint

The mean air concentration in the I-joint has been determined for both submerged and impinging jet impact conditions. The former case corresponds to a jet outlet that is situated underneath the plunge pool water level and, thus, no air is present in the plunge pool. The latter case entrains a lot of free air in the plunge pool and generates a turbulent two-phase shear layer.

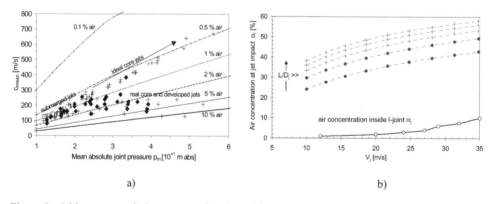

a)   b)

Figure 6. a) Mean wave celerity c_{mean} as a function of the mean pressure value p_m inside the joint. Results are for core (+ sign), developed (♦ sign) and submerged (sign) jet impact conditions. The indicated air volumes are at standard pressure; b) Comparison between the theoretically estimated air concentration at the point of jet impact in the plunge pool and the derived mean air concentration in the I-joint.

Figure 6a summarizes the measurements of wave celerities and, thus, free air concentrations. For submerged jets and ideal core jets, for which theoretically no free air is available at the entry of the rock joint, a certain amount of free air apparently is present inside the joint. The mass of free air changes with the mean pressure at a slope that is steeper than the one that corresponds to a constant mass of free air in the liquid. The mean air content is 1-2 % at low, almost atmospheric mean pressures, while at a mean pressure head of 50 m, the air content decreases down to 0.5 %. This clearly indicates that air is inherently present in the water, and that this quantity can be released inside the joint. Two explanations seem plausible. The air could be present under the form of microbubbles (not visible to the eye) or could be instantaneously released and re-soluted from the liquid during sudden pressure changes inside the joint.

The mean air content for realistic core jets and for developed jets exhibits significant scatter. There is a general tendency towards air contents that are much higher than the ones for ideal core jets or for submerged jets. Air contents of up to 10 % have been observed.

At high pressures, the mean air content inside the joint can fluctuate from 1 to 10 %, i.e. one order of magnitude. These differences in air content are due to the variety of flow conditions in the plunge pool: a turbulent shear layer containing a lot of free air bubbles impacts the joint entrance. The air content in this shear layer, however, continuously changes with time. Assuming that this free air is systematically transferred towards the joint, it explains the observed scatter of the free air content inside the joint.

By defining some average value of the mean air contents observed inside the I-joint, and by relating this value to the theoretical estimations of the corresponding mean air concentration at the point of jet impact in the plunge pool (Figure 4), a direct relationship between the two can be pointed out. This is presented in Figure 6b and shows that some correlation exists. However, only very rough estimates could be made regarding the air content in the plunge pool. It might be convenient to perform instantaneous measurements of both the air content in the plunge pool, close to the rock joint entrance, and the air content inside the rock joint itself.

5.2 The L-joint

Similar to the I-joint, a fundamental frequency exists over the whole joint length (measured at sensors (c), (d) and (d')). Furthermore, a higher harmonic is present in the part of the joint upstream of the 90° bend, i.e. in the I-part (measured at sensors (c) and (d)). For the determination of the mean air content, an assumption has to be made concerning the "n" value of the mode shapes. For both frequencies, n = 0 is assumed (Figure 7). This corresponds to two fundamental sinusoidal mode shapes: one for the whole joint length, and a second one for the length upstream of the bend. For simplicity, the latter will be referred to as the higher harmonic.

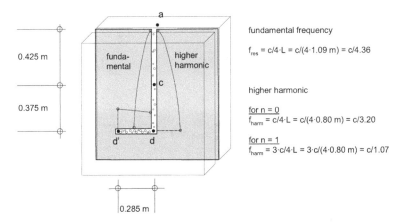

fundamental frequency

$$f_{res} = c/4 \cdot L = c/(4 \cdot 1.09 \text{ m}) = c/4.36$$

higher harmonic

for n = 0
$$f_{harm} = c/4 \cdot L = c/(4 \cdot 0.80 \text{ m}) = c/3.20$$

for n = 1
$$f_{harm} = 3 \cdot c/4 \cdot L = 3 \cdot c/(4 \cdot 0.80 \text{ m}) = c/1.07$$

Figure 7. Scheme of the possible fundamental and higher harmonics of the resonance system of an L-shaped rock joint. For the fundamental frequency, the total joint length is considered. For the higher harmonic, only the part upstream of the 90° bend is used (Bollaert, 2002).

Figure 8 presents the calculated mean air contents for each mode shape separately (Bollaert, 2002). The air contents obtained for the I-joint (Figure 6a) are marked with the symbol ( ). For the higher harmonic, a difference between sensors (c) and (d) has been made. All mode shapes and sensors are presented together in Figure 8d. Based on the fundamental mode in Figure 8a, a mean air content of 5-20 % is observed, values considerably higher than the ones for the I-joint. However, at high jet velocities and mean pressures, the I-joint exhibits air contents that are similar. Based on the higher harmonic, two different results are obtained. The first one is based on sensor (c) situated at the middle of the joint and is presented in Figure 8b. This sensor gives air contents that are in perfect agreement with the ones for the I-joint, i.e. from 1-5 %. The second one is based on sensor (d) situated at the bend and is presented in Figure 8c. For this sensor, surprisingly high celerities have been obtained. These correspond to very low air contents, even lower than the ones for submerged jet impact.

The above analysis depends on the assumptions made about the mode shapes. At first sight, contradictory results are obtained. For the same joint, both low and very high air contents are calculated, which is clearly impossible. This problem can be handled by accounting for the relative importance of each of the mode shapes as a function of the location inside the joint. This is based on the assumption that each distinct part of the joint resonates at its own frequency. Thus, each part has its characteristic air content. For the present L-joint, two parts can be distinguished: upstream and downstream of the 90° bend. An analysis of the spectral contents showed that sensor (c) is significantly influenced by the higher harmonic and that the fundamental mode is hardly present. Sensor (d) exhibits both mode shapes, but the higher harmonic has been displaced compared to sensor (c). This could explain the different air contents at this location. Sensor (d') only shows the fundamental resonance frequency.

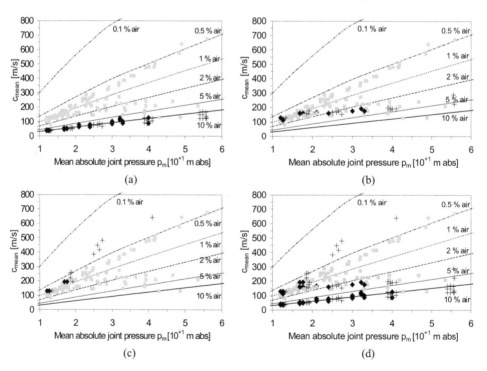

Figure 8. Mean wave celerity inside the L-joint as a function of the mean absolute joint pressure. The data obtained inside the I-joint are marked with a ( ) symbol (Bollaert, 2002).
a) fundamental mode at sensors (c), (d), (d'); b) higher harmonic at sensor (c);
c) higher harmonic at sensor (d) ; d) all mode shapes and sensors together.

146

6 INFLUENCE OF POOL AERATION ON INSTANTANEOUS AIR CONCENTRATION IN ROCK JOINTS

6.1 The I-joint

A summary of results is given in Figure 9 for core, developed and submerged jets, as a function of the instantaneous pressure inside the joint. This instantaneous pressure corresponds to the p_m value as defined in Figure 3.

The air content increases with increasing jet velocity. Hence, like for the mean air content, the instantaneous air content as a function of pressure is largely fluctuating. Furthermore, the quantity of free air available in the joint is clearly not a constant value. The obtained slopes of the celerity-pressure relationships are significantly steeper than the ones for a constant mass of free air. This again implies that a certain amount of air is generated during pressure drops.

It is believed that the present body of experimental evidence points towards air release and resolution as the phenomena that are responsible for the change in free air content. This could at the same time explain the presence of free air under submerged jet impact conditions. The most surprising aspect, however, lies in the instantaneous character of the phenomena. Following the present data, the release and resolution seem to happen quasi instantaneously.

The statement that Henry's law is responsible for the changing air content inside the joint has been verified. For this purpose, a series of curves has been defined for which the air content changes following Henry's law. As outlined in equation (11), a first and simplified relationship can be obtained by expressing the change in quantity of free air as a solubility constant S times a relative pressure drop. The relative pressure drop is expressed as a deviation from a certain equilibrium pressure. This pressure should be the one at which the air bubbles entered the joint. However, as only the relative pressure difference is relevant, the standard atmospheric pressure is chosen here as reference pressure. This equilibrium pressure is than related to a certain air content. As the atmospheric pressure constitutes the lowest possible pressure value inside the joint, all other pressures will exhibit less air. The rate of change in free air content with pressure is determined by the solubility constant S. No time effects are considered.

Three Henry curves are compared with the core jet data from Figure 9a. The results are shown in Figure 9b. Although the Henry curves are more curved than the data, a quite satisfying agreement between data and curves can be observed.

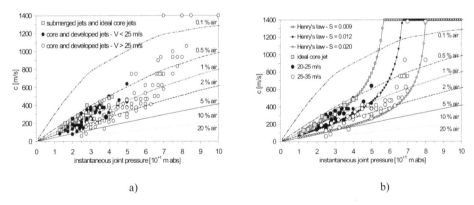

a)   b)

Figure 9.   Instantaneous wave celerity c (m/s) as a function of the instantaneous pressure value p(t) inside the joint: a) for all jet impact conditions; b) for core jet impact conditions. The results are for three different core jets: ideal core impact, real core impact for a velocity of 20-25 m/s and real core impact for a velocity of 25-35 m/s. For each case, comparison is made with an appropriate Henry curve. S stands for the solubility constant as defined in eq. (8).

6.2   The L-joint

For the part situated upstream of the bend, a correlation between sensors (c) and (d) has been made. This procures the celerity-pressure relationship that is valid for this part only. For the part downstream of the bend, the correlation between sensors (d) and (d') results in a second series of celerity-pressure relationships. The results are presented in Figure 10b. The part of the joint between its entrance and the 90° bend is characterized by celerities that are very high, much higher as any previously measured or calculated values. However, sometimes values close to the ones for the I-joint were measured. The part downstream of the bend exhibits celerities that are similar to the I-joint values.

These results have been obtained based on measurements at a 1'000 or 2'000 Hz acquisition rate. The corresponding error in wave celerity is non-negligible. Therefore, the results should only be interpreted as tendencies, and not as absolute values.

Hence, it is concluded that the L-joint is divided into two parts with a completely different air content. The distinction between these parts is made by the presence of the 90° bend. Compared to the I-joint, were a homogeneously distributed air content has been assumed, the present L-joint apparently transfers some of the air in the upstream part towards the horizontal, downstream part. There, the air can get stuck and thus accumulates and forms a cavity. The cushioning effect of this downstream part determines the fundamental resonance mode of the joint. The upstream part adds a second mode shape that is only present upstream of the bend. The closer to the entrance of the joint, the more this second frequency becomes important.

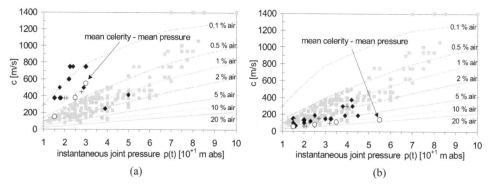

Figure 10.   Instantaneous wave celerity inside the L-joint as a function of the instantaneous joint pressure: a) measured upstream of the bend; b) measured downstream of the bend. The data obtained inside the I-joint are marked with a ( ) symbol.

Furthermore, the mean air contents that were calculated based on the fundamental mode are in reasonable to good agreement with the wave celerity-pressure relationships obtained in Figure 10. With regard to the lack of precision of the used methods, the observed differences are not significant. This is also the case for the mean air contents as derived from the higher harmonic at the bend. On the other hand, the mean air contents calculated based on the higher harmonic at sensor (c) do not correspond to most of the measured celerity-pressure relationships. Although some of the measured points seem to match, it is believed that the higher harmonic at this location has been displaced and, thus, is not representative for the air content.

Finally, it is interesting to verify the above statements by analyzing the pressure signals in the time domain. This is presented in Figure 11, where the signals of the four sensors are presented. Sensor (c) is characterized by a superposition of the fundamental frequency and the higher harmonic. The signal at the bend is dominated by the fundamental frequency, but also exhibits the higher harmonic. However, the peak pressures are completely defined by the fundamental frequency. The signal at the end of the joint only follows the fundamental frequency. These findings confirm the aforementioned assumptions.

148

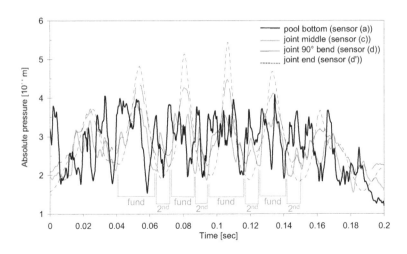

Figure 11. Pressure signals inside the L-joint for sensors (a), (c), (d) and (d').

7 CONCLUSIONS

Based on the experimentally measured water pressures inside rock joints, it has been found that the transient pressure system in these joints is governed by the presence of free air. The amount of free air thereby continuously changes as a function of the pressure in the joint and the amount of free air in the plunge pool. The determination of resonance frequencies in the power spectral content of the pressures inside the joints resulted in time-averaged wave celerities and thus free air contents. These were compared with the theoretically estimated air content at the point of jet impact in the plunge pool and a direct relation between the two has become evident.

Furthermore, direct wave celerity measurements inside the joints defined the instantaneous free air content as a function of the instantaneous pressure value. The shape of these "wave celerity-pressure" relationships is governed by the ideal gas law and Henry's law. Moreover, the instantaneous air content in the plunge pool, which continuously changes with time, dictates their exact position. Hence, an appropriate relationship can hardly be defined.

It is believed that instantaneous measurements of both the air content in the plunge pool, close to the rock joint entrance, and the air content inside the joints, will result in a better correlation between the two.

8 REFERENCES

Bollaert, E. 2001. Spectral density modulation of plunge pool bottom pressures inside rock fissures. *Proceedings of the XXIXth IAHR Congress*, Student Paper Competition, Beijing.

Bollaert, E. & Schleiss, A. 2001. A new approach for better assessment of rock scouring due to high velocity jets at dam spillways. *Proceedings of the 5th ICOLD European Symposium*, Geiranger, Norway.

Ewing, D.J.F. 1980. Allowing for Free Air in Waterhammer Analysis. *Proceedings of the 3rd International Conference on Pressure Surges*, Canterbury, England.

Falvey, H.T.; Ervine, D.A. 1988. Aeration in jets and high velocity flows. *Proceedings of Model-prototype Correlation of Hydraulic Structures*, Colorado, United States.

Martin, C.S. & Padhmanaban, M. 1979). Pressure Pulse Propagation in Two-Component Slug Flow. *Transactions of the ASME*, Journal of Fluids Engineering, Vol. 101, pp. 44-52.

Whiteman, K.J. & Pearsall, I. S. 1959. Reflux Valve and Surge Tests at Kingston Pumping Station. *British Hydromech. Res. Assoc.*, National Engineering Laboratory joint report N°1.

Schweitzer, P.H.; Szebehely, V.G. 1950. Gas evolution in liquids and cavitation. *Journal of Applied Physics*, Vol. 21, pp. 1218-1224.

Wylie, E.B. & Streeter, V.L. 1978). Fluid transients, McGraw-Hill Inc., US.

Time-scale effects, break-up resistance and hydraulic-mechanical interaction

Enguri Arch Dam (Georgia)
(Courtesy of A.J. Schleiss, LCH-EPFL)

The Erodibility Index Method: An overview

G.W. Annandale
Engineering & Hydrosystems Inc., Highlands Ranch, Colorado, USA

ABSTRACT: The paper presents an overview of Annandale's Erodibility Index Method by using a conceptual model to explain the elements of rock scour and summarizing the methodology used to quantify the relative magnitude of the erosive power of water and the relative ability of earth material to resist scour. The scour threshold relating the erosive power of water and the scour resistance of earth material is shown and the method that is used to calculate the extent of scour when using the Erodibility Index Method is explained in a conceptual manner. An approach to quantify of the relative ability of rock to resist scour is detailed in a companion paper to this volume (Annandale 2002b) as is the approach to quantify scour extent (Annandale 2002c).

1 INTRODUCTION

Development of methods that can be used to predict initiation of scour is challenging, and has been the subject of research for many years (e.g. Shields 1936, Hjulstrom 1935, Yang 1973). Most methods have limited application because they either over-simplify the complexity of the hydraulic processes or oversimplify the complexity of factors determining the relative ability of earth material to resist scour.

Successful scour models capture the complexity of the behavior of earth materials as well as the essence of the principal processes that quantify the relative magnitude of the erosive power of water. Assumptions that small-scale processes govern the erosion of earth material often referred to as 'grain-by-grain' removal, can misrepresent actual scour processes because natural earth materials are seldom uniform. The non-uniformity of earth materials is a factor that should be acknowledged when assessing its relative ability to resist erosion. This implies that sole reliance on test results of one parameter, such as undrained shear strength or particle size can potentially lead to incorrect assessment of the relative ability of earth material to resist scour. Experience has shown that large-scale processes often dominate the scour process (e.g. Annandale 1995, Cohen and Von Thun 1994), and that larger units of earth material may scour (removed) prior to grain-by-grain removal from the macro-elements (blocks). This applies to cases of scour of jointed and fractured rock, and to scour of other earth materials such as fissured and slickensided clays. The joints in these materials are often weaker than the crystalline bonds between rock particles or the electro-magnetic bonds created by the Van der Waals forces between clay particles. Failure during the scour process in such cases often proceeds along the discontinuities before the clay or rock blocks themselves (delineated by these discontinuities) break. The Erodibility Index Method (Annandale 1995) empirically incorporates the principal factors that determine the relative ability of earth material to resist scour.

Plucking and cyclic loading introduced by turbulence and amplified by resonance play a dominant role in scouring earth material, especially rock. When water flows over rock (Figure 1) it penetrates the rock discontinuities. The pressure caused by the presence of the water within the discontinuities between the rock blocks is at first equal to hydrostatic pressure, determined by the difference between an elevation within the discontinuity and the elevation of the water surface above the joint. The turbulent flow of water over the rock adds additional forces resulting from the pressure fluctuations at the interface between the rock and the water. These pressure fluctuations are introduced into the discontinuity space and could be amplified by resonance (Bollaert 2001). The net fluctuating pressure forces thus created jack the rock blocks from their positions of rest (Figure 1(a)), and finally dislodge them (Figure 1(b)). Once dislodged, the turbulent water can displace the rock further downstream, provided it has enough power (Figure 1(c)).

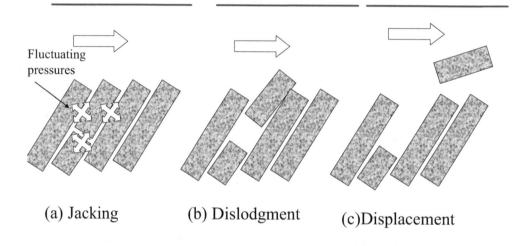

Fluctuating pressures

(a) Jacking   (b) Dislodgment   (c)Displacement

Figure 1. Conceptual model of rock scour processes.

2 MATERIAL RESISTANCE

When considering scour in a complex earth material such as rock its relative ability to resist scour is determined by multiple parameters. Material properties that determine scour resistance of rock include intact material strength, block size, shear strength between blocks of rock, and the relative shape and orientation of the rock blocks. By making use of parameters that represent the relative role of each of these properties to resist scour it is possible to define a geo-mechanical index that quantifies the relative ability of earth material to resist erosion. The geomechanical index used by Annandale (1995) is known as the Erodibility Index (K), which is identical to Kirsten's Excavatability Index (Kirsten 1982):

$$K = M_s \cdot K_b \cdot K_d \cdot J_s \qquad (1)$$

Where M_s = intact material strength number; K_b = block size number; K_d = discontinuity bond shear strength number; and J_s = relative shape and orientation number. Tables and methods to quantify the constituent parameters are presented in Annandale (1995) and Annandale (2002b).

3 EROSIVE POWER OF WATER

The Erodibility Index Method uses stream power, which is equivalent to the rate of energy dissipation in flowing water, to represent the erosive power of water. (These terms are used interchangeably in this paper). By making use of this variable it is possible to quantify the relative magnitude of pressure fluctuations. The hypothesis that the rate of energy dissipation can be used to represent the relative magnitude of pressure fluctuations was demonstrated by Annandale (1995) who analyzed observations pertaining to pressure fluctuations under hydraulic jumps measured by Fiorotto and Rinaldo (1992). The results of the analysis indicated that the standard deviation of pressure fluctuations is directly proportional to the rate of energy dissipation (Figure 2). This finding supports the use of stream power to quantify the relative magnitude of the erosive power of water. Increases in stream power are related to increases in fluctuating pressures, which form the basis of the conceptual model of the erosion process schematized in Figure 1. Methods to quantify the rate of energy dissipation for a variety of flow situations are presented by Annandale (1995), Wittler et al. (1998) and Annandale (1999).

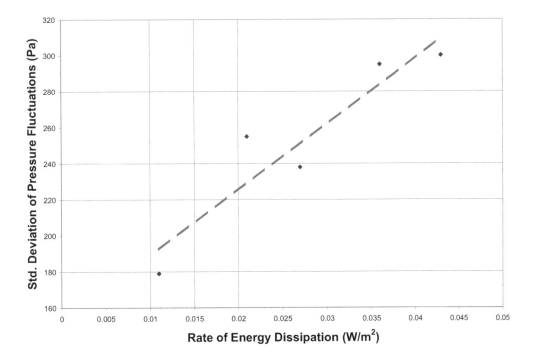

Figure 2. Relationship between the rate of energy dissipation and standard deviation of pressure fluctuations (Annandale 1995).

4 EROSION THRESHOLD

The correlation between stream power (rate of energy dissipation) (P) and a mathematical function ($f(K)$) that represents an earth material's relative ability to resist erosion can, at the erosion threshold, be expressed by the relationship:

$$P = f(K) \tag{2}$$

155

If $P > f(K)$, the erosion threshold is exceeded, and the earth material is expected to erode. Conversely, if $P < f(K)$, the erosion threshold is not exceeded, and the earth material is expected not to erode. The function $f(K)$ represents the Erodibility Index as defined by equation (1).

Annandale (1995) established a relationship between stream power and the Erodibility Index by analyzing published laboratory data pertaining to the erosion threshold of non-cohesive granular material and field data pertaining to the scour of rock, cohesive soils and vegetated soils.

Figure 3 shows the result of the analysis of field data pertaining to scour of cohesive material, vegetated soil, and fractured and jointed rock. Two data types are plotted on the graph, consisting of events where scour occurred and events where scour did not occur. The dotted line indicates the approximate location of the erosion threshold.

Figure 4 contains the results of the analysis of erosion threshold data for non-cohesive granular material ranging from silt to sand, gravel, and cobbles. The results plotted on this graph represent the relationship between stream power and the Erodibility Index at the threshold of erosion. Because the relationship is located at the threshold of erosion, the scatter is less than that on Figure 3.

If all the data is plotted on one graph (Figure 5) the erosion threshold on Figure 4 connects with the erosion threshold on Figure 3 (the dotted line). The dotted line of Figure 3 is not shown on Figure 4, due to scale difficulties, but is located at the lower boundary defined by the set of points in the upper right hand part of the figure that represent scour events. It is shown that the erosion threshold line, as defined by the relationship between stream power and the Erodibility Index, forms a continuous curve for the whole range of earth materials. The earth material represented on Figure 4 ranges from silt (at the lower end of the figure) to hard, intact rock (at the upper end of the figure). The erosion threshold lines presented in Figures 3 through 5 can be used to determine the erodibility of earth materials and to calculate the extent (depth) of scour. The methods used to achieve these objectives are conceptually discussed in what follows.

5 DETERMINATION OF ERODIBILITY

The erodibility of earth materials is determined by plotting the Erodibility Index for a given earth material and the magnitude of the stream power on Figures 3, 4 or 5. If the plotted point is located above the erosion threshold line erosion is expected to occur and if it is located below the threshold line erosion is not expected to occur (Figure 6).

6 DETERMINATION OF EXTENT OF SCOUR

The extent (depth) of scour is determined by comparing the stream power that is **available** to cause scour with the stream power that is **required** to scour the earth material under consideration. The available stream power represents the erosive power of the water discharging over the earth material, whereas the required stream power is the stream power that is required by the earth material for scour to occur. If the available stream power is exactly equal to the required stream power, the material is at the threshold of erosion. In cases where the available stream power exceeds the required stream power, the material will scour. Otherwise, it will remain intact.

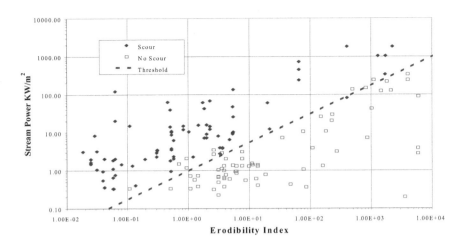

Figure 3. Erosion threshold for upper values of the Erodibility Index (cohesive and vegetated earth material, and rock) (Annandale 1995).

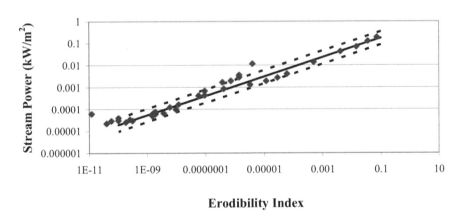

Erodibility Index

Figure 4. Erosion threshold for lower values of the Erodibility Index (non-cohesive granular material) (Annandale 1995).

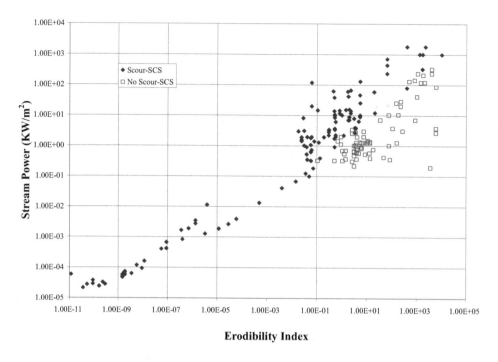

Figure 5. Erosion threshold for an entire range of earth materials, ranging from silt to intact, massive rock – combining Figures 3 and 4 (Annandale 1995).

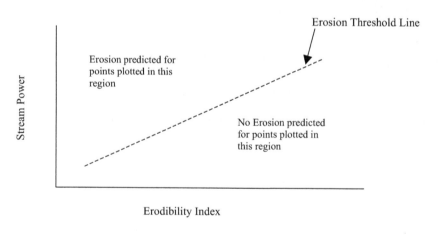

Figure 6. Determination of erodibility of earth materials.

Figure 7 shows how the available and required stream power, both plotted as a function of elevation beneath the riverbed, are compared to determine the extent of scour. Scour will occur when the available stream power exceeds the required stream power. Once the maximum scour elevation is reached the available stream power is less than the required stream power, and scour ceases.

158

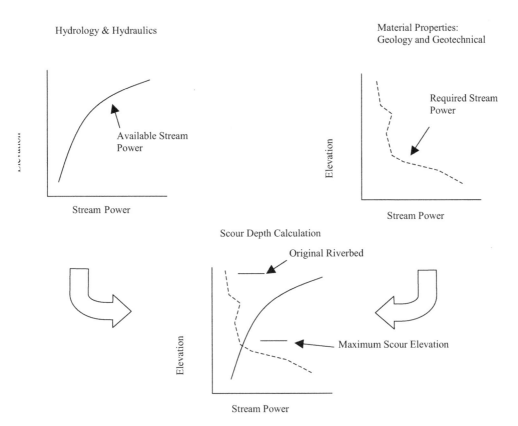

Figure 7. Determination of maximum scour by comparing available and required stream power.

The required stream power is determined by first indexing a geologic core or borehole data. The values of the Erodibility Index thus determined will vary as a function of elevation, dependent on the variation in material properties. Once the index values at various elevations are known, the required stream power is determined from Figure 3 or 5, as conceptually shown in Figure 8. Figure 8 indicates that the stream power required to scour a particular earth material is determined by entering the erosion threshold graph at the abscissa (the Erodibility Index is known) moving vertically to the erosion threshold line, and reading the required stream power on the ordinate. The values of the required stream power are then plotted as a function of elevation for use in the scour calculation.

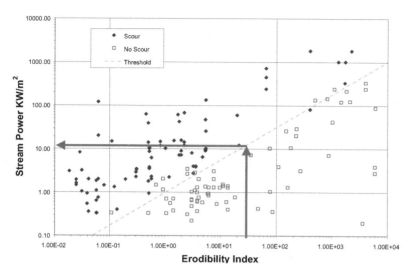

Figure 8. Determination of stream power that is required to scour earth material once the value of the Erodibility Index is known.

7 REFERENCES

Annandale, G.W. 2002b. Quantification of the Relative Ability of Rock to Resist Scour, *International Workshop on Rock Scour*, EPFL, Lausanne, Switzerland, September 25-28, 2002.

Annandale, G.W. 2002c. Quantification of Extent of Scour using the Erodibility Index Method, *International Workshop on Rock Scour*, EPFL, Lausanne, Switzerland, September 25-28, 2002.

Annandale, G.W. 1999. Estimation Of Bridge Pier Scour Using The Erodibility Index Method, in Stream Stability and Scour at Highway Bridges, Richardson, E.V and Lagasse, P.F. (Eds.), *American Society of Civil Engineers, Virginia, USA*.

Annandale, G.W. 1995. Erodibility, *Journal of Hydraulic Research*, Vol. 33, No. 4, pp. 471-494.

Bollaert, E. 2001. Spectral density modulation of plunge pool bottom pressures inside rock fissures. *Proceedings of the XXIXth IAHR Congress*, Student Paper Competition, Beijing.

Cohen, E. and Von Thun, J. L., 1994, Dam Safety Assessment of the Erosion Potential of the Service Spillway at Bartlett Dam, *Proc. International Commission on Large Dams*, Durban, South Africa, pp. 1365-1378.

Hjulstrom, F., 1935, The Morphological Activity of Rivers, *Bulletin of the Geological Institute*, Uppsala, Vol. 25, chapter 3.

Kirsten, H.A.D., 1982, A classification system for excavation in natural materials, *The Civil Engineer in South Africa*, July, pp. 292-308.

Shields, A., 1936, Application of Similarity Principles and Turbulence Research to Bed-Load Movement (in German), *Mitteilungen der Preuss*. Versuchsanstalt fur Wasserbau und Schiffbau, Berlin, No. 26.

Wittler, R.J., Annandale, G.W., Ruff, J.F., Abt, S.R. 1998. Prototype Validation of Erodibility Index for Scour in Granular Media, American Society of Civil Engineers, *Proc. of the 1998 International Water Resources Engineering Conference*, Memphis, Tennessee, August, 1998.

Yang, C. T., 1973, Incipient Motion and Sediment Transport, Journal of the Hydraulics Division, ASCE, Vol. 99, No. HY10, *Proceedings Paper 10067*, pp. 1679-1704.

A physically-based engineering model for the evaluation of the ultimate scour depth due to high-velocity jet impact

E. Bollaert
Laboratory of Hydraulic Constructions, Swiss Federal Institute of Technology, Lausanne, Switzerland

A. Schleiss
Laboratory of Hydraulic Constructions, Swiss Federal Institute of Technology, Lausanne, Switzerland

ABSTRACT: A physically based engineering model has been developed at the Laboratory of Hydraulic Constructions for the evaluation of the ultimate scour depth of a jointed rock mass due to high-velocity jet impact. The model is based on experimental tests and numerical simulations of water pressure fluctuations at plunge pool bottoms and inside artificially created underlying rock joints. The water pressures inside the joints revealed to be of highly transient nature, governed by a cyclic change between high peak pressures and low near-atmospheric pressures. The new engineering model is composed of two sub-models, called the Comprehensive Fracture Mechanics (CFM) model, which uses a simplified Linear Elastic Fracture Mechanics (LEFM) approach to express the erosion resistance of the rock mass, and the Dynamic Impulsion (DI) model, which expresses the ejection of distinct rock blocks from their mass. In the following, these two sub-models are described more in detail.

1 INTRODUCTION

The impact of high-velocity jets onto jointed rock masses creates a progressive erosion of the rock by break-up of the joints and ejection of the so formed blocks. Prediction of this scour phenomenon has been a challenge for engineers since quite a long time. Most of the existing developments are based on model or prototype measurements of scour hole formation as a function of the hydraulic characteristics in question. Only a few methods consider the resistance of the rock mass against this erosion, most of them based on a sort of generalised description.

A new, completely physically based engineering model has been developed at the Laboratory of Hydraulic Constructions of the Swiss Federal Institute of Technology for the prediction of the ultimate scour depth of jointed rock (Bollaert & Schleiss, 2001; Bollaert, 2002; Bollaert et al. 2002). The scour model incorporates two sub-models that express two failure criteria of the rock mass. The first one is called the *Comprehensive Fracture Mechanics* (CFM) model, which determines the ultimate scour depth by expressing the instantaneous or time-dependent crack propagation due to hydrodynamic pressures. The second one is the *Dynamic Impulsion* (DI) model, which describes the ejection of rock blocks from their mass. Each of these sub-models consists of a physical way to describe the destruction of the rock mass (Figure 1, processes 4 and 5). The most appropriate sub-model depends on the geomechanical characteristics of the rock in question, and more specifically on its degree of break-up.

The scour model has been conceived based on three modules: the *falling jet*, the *plunge pool* and the *rock mass*. The latter contains the two aforementioned sub-models that describe failure criteria of rock. In general, emphasis is given on the physical parameters that are necessary to accurately describe the different phenomena. This is, however, performed such that a practicing engineer can easily handle them, which guarantees the comprehensive character of the model without neglecting basic physics behind it.

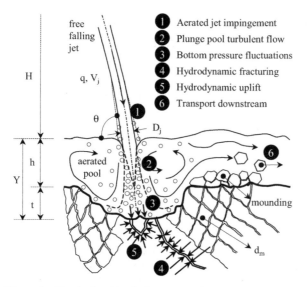

free
falling
jet

q, V_j

H

θ

D_j

1 Aerated jet impingement
2 Plunge pool turbulent flow
3 Bottom pressure fluctuations
4 Hydrodynamic fracturing
5 Hydrodynamic uplift
6 Transport downstream

h

Y

t

aerated
pool

mounding

d_m

Figure 1. Physical-mechanical processes involved in the scouring phenomenon. The present paper focuses on the processes of hydrodynamic fracturing and hydrodynamic uplift, by considering transient pressure wave propagation in the joints of the rock mass.

2 THE MODULES

The scour model consists of three modules: the falling jet, the plunge pool and the rock mass. The modules for the falling jet and for the plunge pool define the hydrodynamic loading that is exerted by the jet at the water-rock interface. The former determines the major characteristics of the jet from its point of issuance at the dam down to the point of impact into the plunge pool. The latter describes the diffusion of the jet through the plunge pool and defines the resulting jet excitation at the water-rock interface. The module for the rock mass has a twofold objective. First of all, it transforms the hydrodynamic loading inside the joints into a critical stress intensity (for closed-end joints) or a net uplift impulsion (for single rock blocks). Secondly, the module of the rock mass defines the basic geomechanical characteristics, relevant for the determination of its resistance. The two relevant failure criteria are *crack propagation* (instantaneous or time-dependent) of closed-end rock joints and *dynamic uplift* or displacement of the rock blocks out of their surrounding mass (for open-end rock joints). For practice, it is recommended to verify both of them, because they are strongly related one to the other. Rock blocks that cannot be ejected from their mass can still be subjected to break-up into smaller pieces (ball-milling effect). These smaller pieces could then be entrained by the flow more easily. On the other hand, even when no further fracturing is possible, the rock mass might already been broken up and capable to be eroded by dynamic uplift.

3 THE FALLING JET MODULE

The module for the falling jet describes the jet trajectory from its point of issuance from the dam down to the impingement of the jet into the plunge pool. Emphasis is given on how the hydraulic and geometric characteristics of the jet are transformed during this trajectory. Hence, the module is based on three parameters that characterize the jet at its point of issuance:

1) velocity at issuance V_i,
2) diameter (or width) at issuance D_i,
3) initial jet turbulence intensity Tu.

The basic idea is to dispose of these initial parameters, or to derive them from the type of outlet structure and the dam reservoir level. The initial jet turbulence intensity Tu defines the spread of the jet during its fall. Furthermore, it will be used to account for low-frequency undulations of the jet. The half angle of outer spread δ_{out} is based on Ervine & Falvey (1987):

$$\frac{\delta_{out}}{X} = 0.38 \cdot Tu \tag{1}$$

with X the longitudinal distance from the point of jet issuance. Typical outer angles of jet spread are 3-4 % for roughly turbulent jets. The corresponding inner angles of jet spread are 0.5-1 %. Although the turbulence intensity Tu is a parameter that reflects the whole range of frequencies of the pressure fluctuations of the jet, it is assumed herein that it also gives an idea about low-frequency fluctuations, i.e. the *compactness* of the jet during its fall. During laboratory tests on near-prototype scale (Bollaert, 2002), it has been found that a significant scatter of the root-mean-square pressure values may exist at the water–rock interface. It is believed that this scatter can be appropriately described by the initial jet turbulence intensity Tu. This is essential because the RMS fluctuations at the surface are generating high peak pressures inside underlying rock joints. The relationship between turbulence intensity and root-mean-square pressure value is presented in Fig. 3. In many cases, Tu is unknown. Under such circumstances, an estimation can be made based on the type of outlet structure:

TYPE OF OUTLET STRUCTURE	TURBULENCE INTENSITY Tu
1. Free overfall	0-3 %
2. Ski-jump outlet	3-5 %
3. Intermediate outlet	3-8 %
4. Bottom outlet	3-8 %

Table 1. Estimation of the initial jet turbulence intensity based on the type of outlet structure.

This classification constitutes a simplification of reality. Tu may largely depend on specific geometric characteristics of the outlet, the flow pattern immediately upstream of the outlet, etc. All these aspects have to be accounted for.

The next step is to define the trajectory of the jet through the atmosphere. A summary of existing equations for different types of outlet structures (free overfall, ski-jump, bottom outlet) has been given by Whittaker & Schleiss (1984). These equations are based on ballistic aspects and drag forces encountered by the jet through the air and will not be further outlined herein. The basic output of these equations is the exact location of the jet impact, the jet trajectory length L and the jet velocity at impact V_j. Knowledge of the jet trajectory length allows determining the contraction of the jet due to gravitational acceleration. This conducts to the jet diameter at impact D_j. This diameter is essential to determine the Y/D_j ratio.

Secondly, the turbulence intensity Tu defines the lateral spread of the jet in equation (1). Superposition of this spread to the initial jet diameter D_i results in the inner and outer limits of the jet diameter D_{in} and D_{out}. The outer diameter of the jet at impact D_{out} will be used to define the maximum zone at the water-rock interface where severe pressure damage might occur. The expressions for D_j and D_{out} are:

$$D_j = D_i \cdot \sqrt{\frac{V_i}{V_j}} \tag{2} \qquad\qquad D_{out} = D_i + 2 \cdot \delta_{out} \cdot L \tag{3}$$

in which V_i and V_j are the jet velocity at issuance and the jet velocity at impact in the pool respectively. The latter is defined by gravitational acceleration as follows:

$$V_j = \sqrt{V_i^2 + 2gZ} \tag{4}$$

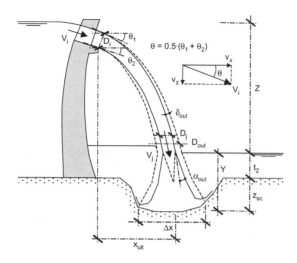

Figure 2. Definition sketch of the parameters of a free overfall jet.

in which Z stands for the vertical fall distance of the jet. The angle of the jet at its point of impact is neglected in the present analysis. For impingement angles that are close to the vertical (70-90°), this effect may be neglected. For smaller impingement angles, it is proposed to use the same hydrodynamic parameters as for vertical impingement, but to redefine the water depth in the pool Y as the exact trajectory length of the jet through the water cushion, and not as the vertical difference between water level and pool bottom. Table 2 summarizes the input and output parameters:

FALLING JET MODULE	
INPUT	OUTPUT
1. Outlet structure: type and geometry	1. Location of jet impact x_{ult}
2. Jet velocity at issuance V_i	2. Jet trajectory length L
3. Jet diameter at issuance D_i	3. Jet impact diameters D_j and D_{out}
4. Initial jet turbulence intensity Tu	4. Jet impact velocity V_j

Table 2. Input and output parameters of the falling jet module.

4 THE PLUNGE POOL MODULE

The second module refers to the hydraulic and geometric characteristics of the plunge pool basin downstream of the dam and defines the hydrodynamic loading at the water-rock interface. The basin may be natural, formed by the shape of the valley and the downstream riverbed, or artificial, by construction of a downstream structure that ensures a minimum tailwater depth at any instant. In case of natural basins, the "tailwater level Y - discharge Q" relationship is calculated in the immediate vicinity of the downstream river. For artificially created basins, it is defined by the control structure. The *water depth in the plunge pool Y* is undoubtedly one of the most important parameters of the scour model. It is defined as the vertical difference between the water level in the basin and the bedrock level. This difference increases with increasing flow discharge Q. Initially, the plunge pool water depth Y equals the tailwater depth "h" as defined in Fig. 1. It is evident that, during scour progression, the water depth Y has to be increased with the depth of the already attained scour "t". Some prototype observations indicate possible mounding at the downstream end of the basin. This mounding results from detached rock blocks that are swept away and that deposit immediately downstream of the basin. This can cause a change of the "tailwater level Y - discharge Q" relationship, but is neglected in the

164

present module. Knowledge of the plunge pool water depth Y together with the output parameters of the jet module (Table 2) allows determining the *ratio of water depth to jet diameter at impact Y/D_j*. This ratio is directly related to the diffusion characteristics of the jet through the pool. However, its definition is purely theoretical and precaution should be taken when applying it to practice. Significant differences may exist due to the appearance of vortices or other surface disturbing effects, which can change the water depth in the pool. No model actually exist that accurately takes into account these effects. Hence, engineering judgment is required on a case-by-case basis. The water depth Y defines the static pressure at the water-rock interface. Although this pressure is often insignificant compared to the dynamic pressures of the impacting jet, it may become relevant when the ultimate scour depth is approached. Near such depths, the diffusion of the jet becomes predominant and eliminates the dynamic action at the bedrock. However, the water depth in the pool often has increased up to several tens of metres. Such static pressures may have an impact on the calculus of crack propagation. The influence on dynamic uplift of single rock blocks is insignificant.

4.1 Definition of root-mean-square pressure fluctuations C'_{pa}

The module for the falling jet pointed out that the turbulence intensity of the jet Tu defines the degree of turbulent pressure fluctuations of the jet at its point of impact in the plunge pool. Hence, the plunge pool module assumes that the root-mean-square value of the pressure fluctuations at the water-rock interface, as expressed by the non-dimensional pressure coefficient C'_{pa} (pressure head divided by $V^2/2g$), depends on both the Y/D_j ratio and the initial turbulence intensity of the jet Tu. This is presented in Fig. 3. The presented data have been measured on an experimental facility (Bollaert, 2002), under the jet's centreline, for jet impact velocities higher than 20 m/s. It is believed that the presented results are exempt of scale effects and, thus, representative for prototype jets. Also, the measured data have been approached by a polynomial regression of third order. This polynomial form has been obtained through curve fitting of the bandwidth of upper data as given by Ervine et al. (1997). The regression coefficient for this curve fitting was equal to 0.99 and yielded the following relationship (Bollaert et al., 2002):

$$C'_{pa} = 0.0022 \cdot \left(\frac{Y}{D_j}\right)^3 - 0.0079 \cdot \left(\frac{Y}{D_j}\right)^2 + 0.0716 \cdot \left(\frac{Y}{D_j}\right) + 0.0583 \qquad (5)$$

The first and the fourth coefficients have been modified in the way indicated at Table 3. This results in four similar-shaped curves but with a different offset. These curves were found to agree with the measured data and can be used up to a Y/D_j ratio of 18-20. For higher ratios, the value that corresponds to a ratio of 18-20 is proposed.

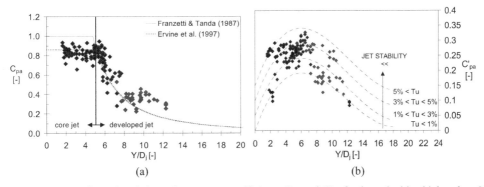

Figure 3. Non-dimensional dynamic pressure coefficients C_{pa} and C'_{pa} for jet velocities higher than 20 m/s: a) mean pressure coefficient. The data corresponds to the best fit curve made by Ervine et al. (1997); b) root-mean-square pressure coefficient. The data are approached by four polynomial regressions of the third order (Bollaert et al., 2002).

It is believed that the curve with the highest root-mean-square values is valid for jets with an undulating or low-frequency turbulent character, or for jets with a turbulence intensity Tu that is higher than 5 %. The curve with the lowest values is applicable to compact jets. These are jets that are smooth-like during their fall, without any possible source of low-frequency instability and with a turbulence intensity that is lower than or equal to 1 %. In between, two other curves have been defined. They are applicable to intermediate stable-unstable or unstable jets.

Tu [%]	a_1	a_2	a_3	a_4	Type of jet
< 1	0.0022	-0.0079	0.0716	0	compact
1-3	0.00215	-0.0079	0.0716	0.050	intermediate
3-5	0.00215	-0.0079	0.0716	0.100	undulating
> 5	0.00215	-0.0079	0.0716	0.150	very undulating

Table 3. Polynomial coefficients and regression coefficient for different turbulence intensities of jets.

The key issue is that the turbulence intensity Tu has been directly related to low-frequency instabilities or undulations of the jet. In fact, Tu reflects the whole range of frequencies of the pressure fluctuations of the jet, and not just the ones that cause undulations and low-frequency instabilities. Nevertheless, it is believed that these effects have a profound impact on the Tu value. The latter might be used to express this phenomenon.

4.2 Definition of mean dynamic pressure coefficient C_{pa}

The non-dimensional mean dynamic pressure coefficient C_{pa} is a function of the Y/D_j ratio. The data are presented in Fig. 4a and in good agreement with the best-fit of data of Ervine et al. (1997). The latter defined C_{pa} as a function of Y/D_j and of the air concentration at impact α_i:

$$C_{pa} = 38.4 \cdot (1 - \alpha_i) \cdot \left(\frac{D_j}{Y}\right)^2 \quad \text{for } Y/D_j > 4\text{-}6 \tag{6}$$

$$C_{pa} = 0.85 \quad \text{for } Y/D_j < 4\text{-}6$$

The air concentration α_i is defined as a function of the volumetric air-to-water ratio β:

$$\alpha_i = \frac{\beta}{1 + \beta} \tag{7}$$

Expressions for β can be found in literature. Following equation (7), the mean dynamic pressure decreases with increasing air content in the plunge pool. This, however, is without accounting for the low-frequency turbulence of the jet, which may have a significant impact on the mean pressure value. As such, on the test facility, the higher mean values were obtained at very high air concentrations, because the jet was more stable under such circumstances. Based on the experimental results (Bollaert, 2002), it is recommended to relate the choice of C_{pa} to the choice of C'_{pa} in the following manner: the higher the chosen curve of root-mean-square values, the lower the choice for the mean pressure value. This is logical considering that turbulent or undulating jets generate high root-mean-square values, but low mean pressures.

4.3 Input and output parameters of the plunge pool module

The input and output parameters of the module for the plunge pool are presented at Table 4. The input parameters correspond to the output of the module for the falling jet. The output parameters will be used as input for the module for the rock mass.

The ultimate scour depth under the jet's centerline can be defined by use of only the first three output parameters. The shape and the extension of the ultimate scour hole strictly need use of five output parameters instead of three. The radial pressure coefficients, however, have been determined for a flat plunge pool bottom. Significant local changes may occur in practice due to a differently shaped bottom. This aspect has never been investigated in detail and is beyond the scope of the project. It should merit attention in further research.

PLUNGE POOL MODULE	
INPUT	OUTPUT
1. Location of jet impact x_{sc}	1. Y/Dj ratio
2. Jet trajectory length L	2. Centreline mean pressure C_{pa}
3. Jet impact diameters D_j and D_{out}	3. Centreline pressure fluctuations C'_{pa}
4. Jet impact velocity V_j	
5. Turbulence intensity Tu	

Table 4.   Input and output parameters of the plunge pool module.

5 THE MODULE OF THE ROCK MASS

The principal parameters of the hydrodynamic loading at the bottom of the plunge pool have been defined in the plunge pool module. This module is used as direct input for determination of the hydrodynamic loading inside underlying open-or closed-end rock joints. The governing parameters are defined hereafter.

5.1   Definition of hydrodynamic loading inside closed-or open-end joints

Four basic parameters describe the hydrodynamic loading inside closed-end or open-end rock joints:
1. maximum dynamic pressure coefficient C^{max}_p
2. characteristic amplitude of pressure cycles Δp_c
3. characteristic frequency of pressure cycles f_c
4. maximum dynamic impulsion coefficient C^{max}_I

The first parameter is relevant to brittle propagation of closed-end rock joints. The second and third parameters are necessary to express time-dependent propagation of closed-end rock joints. The fourth parameter is used to define dynamic uplift of rock blocks formed by open-end rock joints.

5.1.1   Maximum dynamic pressure C^{max}_p in a closed-end rock joint

The maximum dynamic pressure coefficient C^{max}_p is obtained through multiplication of the root-mean-square pressure coefficient C'_{pa} by an amplification factor Γ^+, and by superposing this product to the mean dynamic pressure coefficient C_{pa}. The product of C'_{pa} times Γ^+ results in the coefficient C^+_{pd}. The superposition of C_{pa} and C^+_{pd} is necessary because the amplification of the root-mean-square pressures only influences the fluctuating part of the dynamic pressures. As such, the maximum pressure value is written:

$$P_{max}[Pa] = \gamma \cdot C^{max}_{pd} \cdot \frac{\phi \cdot V_j^2}{2g} = \gamma \cdot \left(C_{pa} + \Gamma^+ \cdot C'_{pa} \right) \cdot \frac{\phi \cdot V_j^2}{2g} \tag{8}$$

in which C_{pa} and C'_{pa} are defined by equations (5) and (6) and γ stands for the specific weight of the water (in N/m³). As a first approach, the ϕ value for non-uniform velocity profiles can be chosen equal to one. The uncertainty lies in the Γ^+ factor. Similar to the root-mean-square pressure values at the bottom of the plunge pool, theoretical curves have been defined for maximum respectively minimum values of the amplification factor Γ^+ as derived from the experiments:

$\Gamma^+ = 4 + 2 \cdot Y/D_j$ for $Y/D_j < 8$
$\Gamma^+ = 20$ for $8 \leq Y/D_j \leq 10$ curve of maximum values
$\Gamma^+ = 40 - 2 \cdot Y/D_j$ for $10 < Y/D_j$

167

$$\Gamma^+ = -8 + 2\cdot Y/D_j \qquad \text{for } Y/Dj < 8$$
$$\Gamma^+ = 8 \qquad\qquad\quad \text{for } 8 \leq Y/D_j \leq 10 \text{ curve of minimum values} \qquad (9)$$
$$\Gamma^+ = 28 - 2\cdot Y/D_j \qquad \text{for } 10 < Y/D_j$$

For jet velocities higher than 20 m/s, the measured amplification factors become more or less independent of the velocity (Fig. 4a). However, considerable scatter is obtained when expressing the values as a function of the Y/D_j ratio (Fig. 4b). For core jet impact, a value of about 4 is systematically obtained at prototype jet velocities. For developed jets, some scatter is observed.

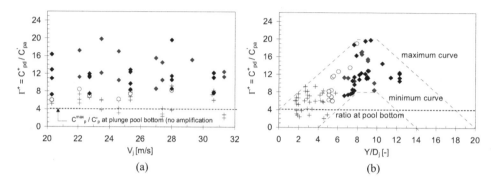

Figure 4. Amplification factor Γ^+ between the maximum pressures inside the joints and the root-mean-square pressures at the water-rock interface: a) as a function of jet velocity V_j; b) as a function of Y/D_j. The data are circumscribed by a maximum curve and a minimum curve.

It may be stated that the scatter in maximum pressures is caused by the exact air concentration inside. Hence, it could be argued that extremely high peak pressures can only be present when no high air content is available or no significant leakage of water occurs out of the joint. This assumption is valid in case of tightly healed rock joints. For such rock joints, the upper bound of the maximum pressure values should be used. For rock joints with several side branches, or joints that are not tightly healed, more air could be present inside. Thus, the lower bound values seem more appropriate.

5.1.2 Characteristic amplitude of pressure cycles Δp_c in a closed-end rock joint

The characteristic amplitude of pressure cycles Δp_c is determined by the characteristic maximum and minimum pressures of the cycles. According to the shape of the pressure spikes, the minimum values are quite constant and always close to the standard atmospheric pressure. The maximum pressures, however, cannot be chosen equal to the C^{max}_p value. The latter is not at all representative for the pressure cycles. In fact, each cycle has a different maximum value. This value can be extremely high, in case of a maximum pressure peak, or can be rather low, for a moderate pressure peak. Hence, an average value of the maximum pressures is needed. This average value has been obtained by counting the number of peaks and by making the average of all the peak values. The number of peaks has been chosen to correspond to the fundamental resonance frequency of each particular test run. As such, the cyclic character of the system is preserved. The results of averaged peak values are presented:

	Core jets	Transitional jets	Developed jets
% of maximum peak pressure	80 %	60 %	40-50 %
N° of times the kinetic energy	1.40	1.30	1.25

Table 5. Definition of the characteristics amplitude of pressure cycles inside rock joints for core, transitional and developed jets.

5.1.3 Characteristic frequency of pressure cycles f_c in a closed-end rock joint

The characteristic frequency of pressure cycles f_c is dictated by the assumption of a $\lambda/4$ – resonator. As such, the air concentration α_i and the total length of the joint L are of significance. An assumption has to be made on the length of the joint. This depends on the interdistance between the different joints. The air content inside the joints can be related to the air content in the plunge pool (Bollaert, 2002). It depends on the mean dynamic pressure coefficient C_{pa} at the bottom of the pool. However, a first hand estimation considers a mean celerity of about 100 m/s at very high mean impact pressures (50 m of water column or more), and a mean celerity of 150-200 m/s at mean impact pressures that are lower than 50 m of water column. The resonance frequency f_{res} is then defined by $f_{res} = c/4 \cdot L$, in which L stands for the total length of the joint. Typical joint lengths are on the order of 0.5 to 2 m. This results in frequencies of 12.5 to 100 Hz, depending on the mean impact pressure.

5.1.4 Maximum dynamic impulsion C^{max}_I in an open-end rock joint

Finally, the maximum dynamic impulsion coefficient C^{max}_I in an open-end rock joint corresponds to a time integration of the net forces on the rock block:

$$I_{\Delta tpulse} = \int_0^{\Delta tpulse} (F_u - F_o - G_b - F_{sh}) \cdot dt = m \cdot V_{\Delta tpulse} \tag{10}$$

For simplicity, a rectangular shaped rock block is assumed, with the sides directed along the horizontal and the vertical respectively. The different forces on such a rock block consist of the upward and downward pressure forces F_u and F_o, the immerged weight of the block G_b and the shear forces along the joints F_{sh}. As a first approach, the latter can be neglected. Furthermore, the immerged weight of the block depends on the shape of the block and can easily be handled mathematically. The problem lies in the determination of the upwards and downwards directed pressure forces. These forces are generated by the pressure fields over and under the block. The pressure field over the block is governed by turbulent shear layer impact. The pressure field under the block corresponds to the transient pressure field that is generated inside open-end rock joints.

The different forces are considered independent of the progressive movement of the rock block. This is a simplification of reality. For example, the shear forces can considerably increase due to a change in block position. On the other hand, the pressure forces under the block could decrease due to the cavity that is formed under the block once the latter starts moving. These effects, however, are hard to formulate. The pressure drop due to cavity formation under the block is not known. The shear forces along the joints depend on the points of contact between the blocks and on the in-situ horizontal stress field of the rock mass. Therefore, some assumptions have to be made. First of all, the transient pressures under the block are believed to be independent of the movement of the block. This seems plausible when the net impulsion consists of a high peak pressure during a small time interval. For such an impulsion, the block has not enough time to move during the initial phase of the impulsion and, hence, cannot influence the subsequent part of the impulsion. For impulsions that consist of a low pressure during a relatively long time interval, this becomes less evident. However, it is a priori not excluded that the water can progressively fill up the formed cavity during several consecutive impulsions.

Hence, the first step is to define the instantaneous differences in pressure forces over and under the block. For periods Δt during which positive differences exist, the time integral has been taken over these periods. This results in net impulsions I. A maximum net impulsion I^{max} can so be defined. This has been done systematically for all test runs. Similar to the pressure coefficients, the maximum net impulsion I^{max} has to be made non-dimensional. This is done by defining the impulsion as the product of a net force times a time period. The net force is firstly transformed into a pressure. This means that the problem is looked at for a unitary surface of the block ($1 m^2$ according to the units). This pressure can then be made non-dimensional by dividing it by the incoming kinetic energy $\phi \cdot V^2/2g$. This results in a net uplift pressure coefficient C_{up}.

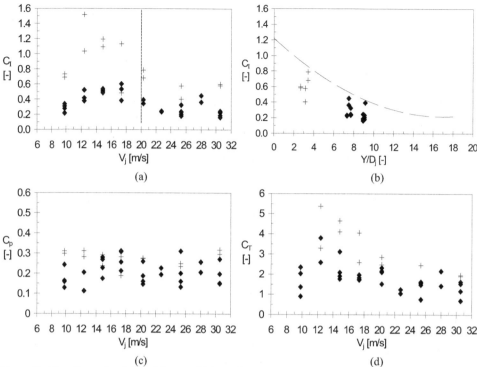

Figure 5. Non-dimensional impulsion coefficients for pressures inside open-end rock joints: a) C_I as a function of V_j; b) C_I as a function of Y/D_j; c) C_p as a function of V_j and d) C_T as a function of V_j.

The time period is non-dimensionalized by the characteristic period for pressure waves inside rock joints, i.e. $T = 2L/c$, in which L stands for the total length of the joint and c is the mean wave celerity. This conducts to a time coefficient T_{up}. Hence, the non-dimensional impulsion coefficient C_I is defined by the product $C_{up} \cdot T_{up} = V^2 L/gc$ [m·s] and is presented in Fig. 5. The net maximum impulsion I^{max} is finally obtained by multiplication of the value for C_I by $V^2 L/gc$. Fig. 5a shows that, for jet velocities higher than 20 m/s, a quite constant value for C_I is observed. This is confirmed in Fig. 5b, where the coefficient C_I is presented as a function of the Y/D_j ratio. The observable scatter is quite low. For core jets, a value of 0.6-0.8 seems plausible. For developed jets, the values are situated between 0.2 and 0.5. These results are valid for test runs of approximately 1 min. It can be argued that much longer time periods are needed to measure the real maximum value. This statement, however, has been derived from scale-model studies. Its relevance to near-prototype model studies has not yet been pointed out. For practice, it is proposed to use the maximum measured values of Fig. 5b as C^{max}_I. This leads to the following definition:

$$C_I = 0.0035 \cdot \left(\frac{Y}{D_j} \right)^2 - 0.119 \cdot \left(\frac{Y}{D_j} \right) + 1.22 \tag{11}$$

The character of the impulsions is determined by the average pressure value, by the ratio of maximum to average pressure value and by the time duration of the pulse. As such, the character cannot be deduced from the C_I coefficients alone. For example, a low pressure that holds for a long time can exhibit the same impulsion as a short-lived pressure peak. However, the character can be determined by the C_{up} and T_{up} values separately. At low jet velocities, the significant increase in C_I values is due to an increase of the time duration of the pulse rather than an increase of its average pressure value. This is valid for both core and developed jets.

5.2 Definition of failure criteria for closed-end rock joints

The cyclic character of the hydrodynamic loading generated by the impact of a high-velocity jet on a closed-end rock joint makes it possible to describe rock joint propagation by fatigue effects at the tip of the joint. Linear elastic fracture mechanics (LEFM) models can handle both static and dynamic loadings and resistances assuming a perfectly linear elastic, homogeneous and isotropic material. These models become quite complicated when accounting for all the relevant parameters. Therefore, a simplified application is presented here. The application represents a practical engineering approach of the underlying theory and attempts to describe the parameters of influence such that an engineer can easily handle them. In other words, the model still accounts for the correct physical tendencies but has a comprehensive character.

It is called the *Comprehensive Fracture Mechanics* (CFM) *Model* and is applied in the present section to intermittently jointed rock. Hydrodynamic loading in mode-I (pure tensile) is described by the stress intensity factor K_I. The corresponding resistance of the material is expressed by the fracture toughness value K_{Ic}. The problem lies in a comprehensive and physically correct implementation of the complex and dynamic situation encountered in rock joints. Crack propagation distinguishes between brittle (or instantaneous) crack propagation and time-dependent crack propagation. The former happens for a stress intensity factor that is equal to or higher than the fracture toughness of the material. It occurs for both static and dynamic loadings. The latter is valid when the applied static or dynamic loading is inferior to the material's resistance. Cracks are propagated by fatigue. Failure by fatigue depends on the number and the amplitude of the load cycles. The fracture mechanics implementation of the hydrodynamic loading consists of a transformation of the water pressures s_{water} into rock stresses at the crack tip. These stresses are characterized by the stress intensity factor K_I, written as follows:

$$K_I = \sigma_{water} \cdot f \cdot \sqrt{\pi a} \tag{12}$$

in which K_I is expressed in MPa$\sqrt{m}$ and the water pressures in MPa. The implementation makes use of the following simplifying assumptions: 1) the dynamic character of the loading has no influence; 2) the pressure distribution inside the joints is constant; 3) only simple geometrical configurations of rock joints are considered; 4) the joint surfaces are planar.

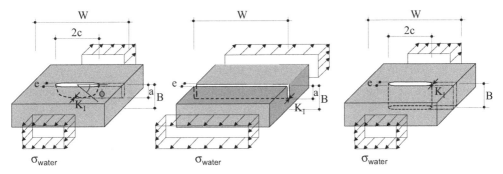

Figure 6. Proposed framework for the basic geometrical configurations of intermittently jointed rock: a) semi-elliptical (EL) joint; b) single edge (SE) joint; c) center-cracked (CC) joint. The water pressures are applied from outside the joints.

The boundary correction factor *f* is a function of the type of crack. Figure 6 presents the three basic situations for intermittently jointed rock. The water pressure in the joints is applied from outside the rock. No geometries with multiple joints are considered. The choice of the most relevant geometry depends on the type and the degree of jointing of the rock in question. The first crack is semi-elliptical or semi-circular in shape and, with regard to the laterally applied stress σ_{water}, partially sustained by the surrounding rock mass in two directions. As such, it is the geometry with the highest possible support of surrounding rock. Corresponding stress intensity factors should be used in case of low to moderately jointed rock. The second crack is single edge notched and of two-dimensional nature. Support from the surrounding rock mass is only exerted perpendicular to the plane of the notch and, as a result, stress intensity factors will be substantially higher than for the first case. Thus, it is more appropriate for moderately to highly jointed rock. The third geometry is center-cracked throughout the rock. Similar to the single edge notch, only one-sided rock support can be accounted for. This support, however, should be slightly higher as for the single edge notch. The second and third configurations correspond to a partial destruction of the first one. As such, they both are more sensible to stresses and have to be used for moderately to highly jointed rock. A summary of f values is presented in Figure 7.

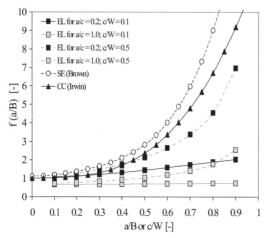

Figure 7. Comparison of different boundary correction factors f.

The fracture toughness is depending on a vast range of parameters. In the following, it is related to the mineralogical type of rock and to the tensile strength T or the unconfined compressive strength UCS. Furthermore, corrections are made to account for the effects of the loading rate and the in-situ stress field of the rock mass. The corrected fracture toughness is here called the in-situ fracture toughness $K_{I,ins}$. The following formulae are proposed:

$$K_{I\,ins,\,T} = (0.105 \text{ to } 0.132) \cdot T + (0.054 \cdot s_c) + 0.5276 \qquad (13)$$
$$K_{I\,ins,\,UCS} = (0.008 \text{ to } 0.010) \cdot UCS + (0.054 \cdot s_c) + 0.42 \qquad (14)$$

in which T, UCS and σ_c are expressed in MPa. Instantaneous or brittle crack propagation will occur if the following expression is valid:

$$K_I \geq K_{I,ins} \qquad (15)$$

If this is not the case, crack propagation still occurs within a certain time interval. This is expressed by an equation of the type that was originally proposed to describe fatigue crack growth in metals:

$$\frac{da}{dN} = C \cdot (\Delta K_I)^m \qquad (16)$$

in which a is the crack length and N the number of cycles. C and m are rock material parameters that can be determined by experiments and ΔK_I is the difference of maximum and minimum stress intensity factors at the crack tip. To implement time-dependent crack propagation into a comprehensive engineering model, the parameters m and C have to be determined. The parameters m and C have been summarized at Table 7 for a range of rocks.

5.2.1	Type of rock	Exponent m	Coefficient C
	Arkansas novaculite	8.5	1.0E-8
	Mojave quartzite	10.2-12.9	3.0E-10
	Tennessee sandstone	4.8	4.0E-7
	Solenhofen limestone	8.8-9.5	1.1E-8
	Falerans micrite	8.8	1.1E-8
	Tennessee marble	3.1	2.0E-6
	Westerley granite	11.8-11.9	8.0E-10
	Yugawara andesite	8.8	1.1E-8
	Black gabbro	9.9-12.2	4.0E-9 - 5.0E-10
	Ralston basalt	8.2	1.8E-8
	Whin Sill dolerite	9.9	4.0E-9

Table 7. Fatigue exponent m and coefficient C for diffferent types of rock.

5.3 Definition of failure criteria for open-end rock joints

The *Dynamic Impulsion* model is based on the maximum dynamic pressure impulsion C^{max}_{I}, as defined before, and on the geometrical characteristics of a rock block that is representative for the rock mass. The geometrical characteristics, together with the type of rock, define the immerged weight of the block. Furthermore, the shear forces are determined by the horizontal in-situ stress field. In a first approach, these can be neglected by assuming that progressive dislodgment and opening of the joints occurred during the break-up phase of the rock mass.

A definition sketch of possible geometrical situations is provided in Fig. 8. The rock blocks are considered squared in the horizontal plane. This means that the analysis is two-dimensional in the x-z plane. The side length in the x-direction is called x_b, while the vertical side is of length z_b. As such, the total length of the joint that passes underneath a block is equal to $L = x_b + 2 \cdot z_b$. The thickness of the joint is neglected. The two-dimensional character implies that the forces on the rock blocks are considered constant in the y-direction. For simplicity, the pressures will be considered equal to the forces, i.e. acting on a surface of 1 m^2. Assuming a bandwidth of 1 m in the y-direction, this means that the analysis is performed per m of length in the x-direction. The immerged weight of the block G_b can be defined as follows:

$$G_b = \forall \cdot (\gamma_r - \gamma_w) = x_b^2 \cdot z_b \cdot (\gamma_r - \gamma_w) \qquad (17)$$

in which γ_r and γ_w stand for the specific weight of the rock respectively the water. For a horizontal surface of 1 m^2, the weight force becomes linearly dependent on the height of the block. For a bandwidth of 1 m, the problem becomes dependent on the form factor z_b/x_b of the rock block. Three cases of form factors can be considered: $z_b/x_b > 1$, $z_b/x_b = 1$ and $z_b/x_b < 1$ (Fig. 8).

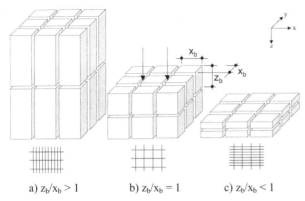

a) $z_b/x_b > 1$ b) $z_b/x_b = 1$ c) $z_b/x_b < 1$

Figure 8. Three cases of form factors of a characteristic rock block. The influence of the y-direction is neglected in the present analysis.

These three cases have the same horizontal surface and, thus, the same net uplift pressure. However, their immerged weight per m² is completely different. It is obvious that block a) will be much more difficult to eject than block c). Moreover, the shear forces obviously will also be higher in case a) than in case c). As a result, knowledge about the joint sets and the distance between the joints is of crucial importance to define the ultimate scour depth based on dynamic uplift of rock blocks. The weight and shear forces are considered constant in time. This means that they can be simply subtracted from the net impulsion calculated in eq. (10) by multiplying the forces by the time period of the net pressure pulse. This net impulsion is then set equal to the product of the mass of the block times the velocity, which results in the maximum velocity that could be given to the mass of rock. This velocity or kinetic energy is transformed into potential energy by ejection of the block. Equation (18) expresses the displacement h_{up} of the block as a function of this velocity:

$$V_{\Delta tpulse} = \sqrt{2 \cdot g \cdot h_{up}} \tag{18}$$

The uplift is proportional to the square of the velocity. This again indicates the importance of the vertical height z_b of the rock block. The displacement that is necessary to eject a rock block from its matrix is hard to define. It depends on the degree of interlocking of the blocks, which depends on the in-situ stress field of the rock mass. As a first approach, a very tightly jointed rock mass will need a displacement that is equal to or higher than the height z of the block. Less tightly jointed rock will probably be uplifted more easily. The necessary displacement is a model parameter that needs to be calibrated.

6 CONCLUSIONS

The two major physical processes that are responsible for rock scour due to high-velocity jet impact have been expressed by means of two physically-based models: the *Comprehensive Fracture Mechanics* (CFM) model, which handles instantaneous or time-dependent crack propagation of closed-end rock joints and the *Dynamic Impulsion* (DI) model, which deals with dynamic ejection of rock blocks from their surrounding mass.

The hydrodynamic loading inside these joints has been quantified by means of non-dimensional pressure and impulsion coefficients. The corresponding pressure fluctuations depend on the degree of turbulence of the falling jet and the diffusion of the jet through the plunge pool water depth. The resistance of the rock mass against scouring has been expressed by appropriate failure criteria. These criteria can be applied with a minimum of knowledge on the geomechanical characteristics of the rock mass.

This direct coupling of hydrodynamic and geomechanic aspects is the base of the present new scour evaluation method. Its application is such that a practicing engineer can easily understand and handle the parameters in question. An application example can be found in Bollaert et al. (2002) and gives a first idea about the sensibility of these parameters.

7 REFERENCES

Bollaert, E. 2002. Transient water pressures in joints and formation of scour due to high-velocity jet impact, *PhD Thesis Report*. Laboratory of Hydraulic Constructions, EPFL, Lausanne, Switzerland.

Bollaert, E. & Schleiss, A. 2001. A new approach for better assessment of rock scouring due to high velocity jets at dam spillways. *Proceedings of the 5th ICOLD European Symposium*. Geiranger, Norway.

Bollaert, E., Falvey, H.T., Schleiss, A. 2002. Turbulent jet impingement in plunge pools: the particular characteristics of a near-prototype physical model study. *Accepted for presentation at the conference Riverflow 2002*. Louvain-la-Neuve, Belgium.

Ervine, D.A.; Falvey, H.R. 1987. Behavior of turbulent jets in the atmosphere and in plunge pools. *Proceedings of the Institution of Civil Engineers*. Part 2, Vol. 83, pp. 295-314.

Ervine, D.A.; Falvey, H.R.; Withers, W. 1997. Pressure fluctuations on plunge pool floors. *Journal of Hydraulic Research*, IAHR, Vol. 35, N°2.

Whittaker, J.; Schleiss, A. 1984. Scour related to energy dissipators for high head structures. Zürich.

Geotechnical aspects of rock erosion

J.H. May & J.L. Wibowo
US Army Engineer Research and Development Center, Vicksburg, MS

C. C. Mathewson
Texas A & M University, College Station, TX

ABSTRACT: Severe erosion in unlined spillway channels at large Corps of Engineer reservoirs in the 1980s and 1990s caused the Corps to initiate research to predict erosion damage in unlined bedrock spillways. The purpose of this research was to specifically study the erosion phenomena with respect to the combined effects of the geologic and hydrodynamic controls. The research revealed that most of the severe erosion in the unlined spillways was caused by knickpoint (overfall) erosion where large blocks of material had been undercut causing mass failures. It was concluded that the potential for severe erosion was controlled by the geology at the overfall and point of impact, the velocity of the eroding water, and the negative pressures underneath the overfall. The factors that influence geologic resistance to hydraulic erosion which were identified in this research effort can be applied at other sites where rock erosion is a problem.

1 INTRODUCTION

All geologic materials are erodible under the influence of moving water. If given sufficient quantity, and duration of flow, water will erode these materials to a base level where relative stability will occur. In regard to dams, spillways and associated structures, the rate of erosion is the critical factor to consider in the initial design, or in any required remedial actions. At existing structures, material that will rapidly erode must be recognized. Engineering properties such as cohesiveness, hardness, compressive strength, rock quality designation, and permeability are some of the important parameters in considering how fast a geologic material will erode.

Several factors prompted the US Army Corps of Engineers (CE) to include the problem of rock erosion in unlined spillway channels as a work unit in the Repair, Evaluation, Maintenance, and Rehabilitation (REMR) Research Program conducted by the US Army Engineer Waterways Experiment Station (WES) during FY 85-89. REMR was a comprehensive investigation of the problems associated with the maintenance and preservation of Civil Works structures constructed and operated by the CE. Experience spanning two decades demonstrates that severe erosion of rock and soils in unlined emergency/auxiliary spillway channels may cause undermining or failure of spillway structures and catastrophic release of reservoir waters. Significant erosion-induced damage is well-documented in unlined emergency/auxiliary spillway channels at flood control and water-storage projects built and managed by the CE, other Federal Agencies, state, and local interests (Cameron et al. 1986, 1988a, 1988b).

The prediction of initiation, rate, and intensity of channel erosion during spillway overflow is far from being a precise science; and effective, cost-efficient, engineered solutions regarding erosion prevention and remediation are often difficult to perceive, justify, and enact. The complex interrelationship between site-specific geological features and hydrodynamic factors leading to unlined spillway channel erosion was, (and to a certain extent still is), poorly understood. The overall objective of the REMR research was to develop procedures for predicting, detecting, preventing, and repairing rock erosion in unlined spillway channels. The purpose of this paper is to summarize the findings of the REMR research and to focus on the geotechnical aspects of predicting rock erosion that can be applied to sites where rock erosion could be a problem.

2 SPILLWAY EROSION STUDIES

The American Society of Civil Engineers and the United States Committee on Large Dams document spillway erosion by summarizing large dam failures and incidents dating to the late 1800's for United States dams (ASCE/USCLD 1975, 1988). Where possible, factors such as erosion involving only concrete, flow damaging only the controlling gates, or dam overtopping by excessive discharge are listed. However, a considerable number of incidents involve erosion of soil and rock. See Cato (1991) for a discussion of historical data.

The Repair, Evaluation, Maintenance, and Rehabilitation (REMR) Research Program spillway erosion study began in 1984 as a result of substantial spillway erosion at the CE Grapevine Dam, TX, in 1981, and the CE Saylorville Dam, IA, in 1984 (USACE 1962, 1984). Flow magnitudes represented small percentages of the Probable Maximum Flood (PMF), but large rock volumes were eroded. More importantly, it was felt that larger flows would have produced spillway breaches.

The REMR study consisted of a reconnaissance stage assessing the problem magnitude and a research stage addressing specific issues. The reconnaissance study, performed by a multidisciplinary team based out of the US Army Engineer WES in Vicksburg, MS, entailed contacting every CE Division to identify historic spillway flows. Efforts were made to visit each site, evaluate geologic materials in the spillway channel, and write up an event case history. Further research resulted in the following conclusions (Cameron et al. 1988a):
- Structural and stratigraphic discontinuities play a major role in the erosion of rock by changing the erosion resistance of the bed material and forming channel gradient changes.
- It is possible to rank erosion at sites by comparing volumes of material removed.
- Headward knickpoint migration can be exacerbated by negative pressures pulling the turbulent forces of the nappe toward the natural materials in the headcut.

The Soil Conservation Service (SCS) Emergency Spillway Flow Study Task Group (ESFSTG) evaluated the performance of more than 75 spillway flows to refine design criteria and guide repair of eroded sites. The SCS data provided a large percentage of the overall REMR research data. The SCS dams were large enough for valid comparisons to CE dams. Erosion severity in the data base varies from no damage to one complete breach. Observations show that most eroded material consists of soil placed on the exit channel floor and residual soil on the natural hillslope. Involvement with CE and SCS spillway studies led Cato (1991) to probe material performance case histories. Geometric and hydraulic effects on erosion processes were analyzed by statistically comparing erosion damage to geometric and hydraulic variables for 16 sites; portions of this work are summarized in Cameron et al. (1988a). The analysis resulted in a method to classify erosion as dominantly downcutting, transition or backcutting.

This initial study guided succeeding analyses to consider all three general variables (geometry, hydraulics, and geology). It was found that site geometry is a more critical factor in controlling the initiation of erosion than flow hydraulics and that site geology appears to serve as the dominant erosion control factor.

Floods in 1993 caused severe damage in Tuttle Creek, Kansas, and in Saylorville, Iowa, as well as in Painted Rock, Arizona. As a result of these floods the research effort on Unlined Spillway Erosion has been continued by the US Army Engineer Research Development Center (ERDC) as an extension to REMR. Additional unlined spillway erosion case histories were analyzed, and more recent erosion prediction technology (Temple, 1994, and Annandale, 1995) has been carefully studied (Wibowo and Murphy, 2002).

3 KNICKPOINT/OVERFALL EROSION

Knickpoint migration or headcutting is the most severe form of structure-threatening erosion in emergency/auxiliary spillway channels. It is also the most unpredictable form of erosion because it is controlled mainly by the geology at a particular site. Erosion studies analyzed geometric and hydraulic parameters in respect to severity of erosion (Cameron et al. 1988b). The absence of a statistical correlation among the variable parameters is ascribed to different geological conditions at the data base sites. Structural and stratigraphic discontinuities were not included in the regression analyses.

Studies dealing with the erosion caused by free falling trajectory jets may be more realistic to describe the turbulent and dynamic end of the "erosion continuum" which ranges from erosion of individual grains to movement of massive blocks of rock. Vieux (1986) of the Soil Conservation Service studied plunge pool erosion in cohesive soils dams in Kansas, where scouring at horizontal pipe outlets has occurred. A formula based on unconfined compressive strength was used to predict the equilibrium scour pool length to within 5 percent of measured length

Mason (1984) reviewed the case histories of scour development selected prototypes including: the Tarbela Dam in Pakistan; Alder and Nacimiento Dam in the United States; Picote Dam in Portugal and Grand Rapids Dam in Canada; and Kariba Dam in Zimbabwe. Mason pointed out the difficulties in quantitatively predicting the erosion characteristics of free-trajectory jets. Unacceptable scour is by no means limited to soft rocks such as shales, sandstones and limestones but occurred in competent igneous and metamorphic rocks such as andesite, granite and gneiss. He also discussed hydraulic model testing and said that in every case he had studied the Froude law was used as the scaling law.

Mason developed formulae for calculating the depths of erosion under free jets for both models and prototypes. He noted that the most common problem in modeling scour is representing the geology of the site at a model scale. A common approach is to examine the rock on site and to estimate the size of the blocks that will result from the joint and fissure patterns. The rock in the model is then represented by an equivalent size of gravel. The gravel can be left as noncohesive for a worst case condition or mixed with various percentages of clay, cement, chalk powder and water to add cohesion.

A rational approach was presented by Spurr (1985) for estimating the scour downstream of large dams by taking into account the mean surplus jet energy in relation to the geology and estimated flow durations. According to Spurr, the scour depth, D_t, at any time, t, resulting from a submerged jet eroding bedrock is a function of the jet energy, E_a available at the surface of the bedrock and the rock's capacity, E_b, to absorb or deflect the erosive forces, such that:

$$D_t = f(E_a - E_b - E_x) \qquad (1)$$

where E_x is the jet energy deflected by the bedrock at time, t. Spurr presented the most important geological considerations as:

- The rock mass's resistance to hydrofracture, as governed by its tensile properties, its degree of fracturing, faulting, and bedding plane spacings; and
- The bedrock's resistance to the erosive shear forces exerted on its surface by the action of the jet, as governed by its cohesive strength.

Spurr applied the empirical formula for scour depth (Equation 1) developed by Mason (1984). This formula reproduced the known scour depth at the prototype reference sites and effectively calibrated Spurr's model. Blaisdell (1983) and Blaisdell and Anderson (1984) also studied ways to predict scour at cantilevered pipe outlets. Analyses of laboratory and field data indicated that the maximum depth of scour, in terms of the pipe diameter, D, and height of overall Z, occurs when the ratio, Z/D, approximately equals 5. Reinius (1986) measured the water pressures around rock blocks in a hydraulic flume. His study was based on the assumption that the rock has cracks in several directions and that water can enter the cracks causing pressure to build within them. If the uplift force is not offset by the weight of the block, and no other forces in the joints stop movement, the block will be lifted and carried away.

During the study of numerous case histories it was noted that various types of erosion could occur at the same location within a given spillway during a single flood event. Tractive force scour is usually taken into consideration in the design of emergency/auxiliary spillways. The Soil Conservation Service has developed excellent criteria for the design of grass-lined spillways (Temple 1980, 1982, 1983. 1984, 1986,). Temple and Hanson (1993) defined the process of erosion in three phases: the stripping of the vegetal cover, the development of a knickpoint, and the advancement of the knickpoint. In analyzing the advancement of a knickpoint, Moore et al. (1994) related the energy of hydraulic attack and the material erosion index (Kirsten, 1988) using US Department of Agriculture (USDA) case histories, and developed a spillway erosion threshold line which was later implemented in the Sites computer program for spillway erosion analysis. Annandale (1995) developed a more rigorous approach in relating the rate of energy dissipation and the material erosion index. Using the USDA spillway erosion database and some additional case histories from US Bureau of Reclamation (USBR) and South Africa he developed a threshold line for a more general case of erosion. ERDC is continuing the effort that was done by REMR in studying the unlined spillway erosion problem. Adopting the material erosion index used by Temple and Annandale, ERDC has collected more spillway erosion case histories and improved the spillway erosion threshold line (Wibowo and Murphy, 2002).

3.1 Hydrodynamic mechanism

The phenomenon of undercutting has been documented for years, but the actual mechanics involved were largely unknown. During the REMR flume studies conducted in this research the actual sequence of events, geologic conditions, geometries, and velocities necessary to cause undercutting were actually reproduced and understood.

The flume studies have demonstrated that the mechanics of knickpoint migration or head-cutting are controlled by the geometry of the knickpoint and the velocity of the flow. The geometry of the knickpoint is in turn governed by the geology at the specific site. Therefore, geology, in conjunction with topography, plays a key role in the initial location of a knickpoint. Because the knickpoint plays such an important role in emergency/auxiliary spillway erosion, it is necessary to understand what is taking place at the knickpoint from a hydraulics perspective. The review of literature directly relevant to knickpoint formation and growth showed that the actual mechanics, which controlled the process, was largely unknown. It was noted during the REMR research that a hydraulic "drop structure" is geometrically and hydraulically very similar to a knickpoint. There is a great deal of energy loss in the area where the falling jet strikes the floor of the channel. The amount of energy loss has been determined experimentally by Moore (Moore, 1943).

Based on the concept that drop structures are designed to prevent scour and headward erosion, it seemed appropriate to assume that knickpoints which happen to meet the same geometric and hydraulic standards would be much less likely to have severe erosion potential. Conceptually, the overall research effort was geared toward creating a model for a worst-case scenario for headward migration of a knickpoint and determining the role of stratigraphic variation on erosion using rock simulants of gravel, gelatin, plexiglas and sodium silicate (May, 1989).

In order for the calculations for the jet impact angle, **2**, to be valid, the area underneath the waterfall must be vented to the atmosphere. If this area is not vented a pressure differential will occur which will draw the waterfall inward against the vertical face and cause severe erosion. The unvented condition was used to study the worst-case condition for undercutting. In unvented conditions the waterfall can be drawn underneath an erodible layer at an angle greater than 90°. The tail water depth is critical in controlling the stresses that are acting on the area underneath an overfall. For a particular set of knickpoint or drop structure conditions, stresses will be transmitted to the channel floor in proportion to the depth of the tail water. The tail water was lowered as much as possible during the REMR modeling efforts so that the energy of the falling jet was at a maximum.

For headcutting to occur the reverse roller portion of the overfall must be in close proximity to the vertical face of the knickpoint. The reverse roller can come in contact with the vertical face in one of several ways:

– a vented knickpoint in which the ratio of height of fall, Z, to depth of flow, Y, is greater than 8/1 (Fig. 1) or;
– an unvented knickpoint in which negative pressures hold the flow against the vertical face. In the vented case, erosion is controlled by the flow conditions and the knickpoint geometry.

In the unvented case the pressure differential can cause the water to be drawn against the vertical face regardless of the Z/Y ratio. For any given set of geometric and velocity conditions the

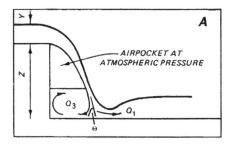

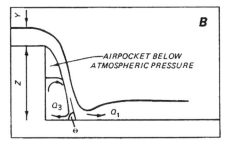

Figure 1. Schematic diagram showing vented (A) and unvented (B) conditions at a knickpoint. Y equals depth of flow, Z equals height of overfall and 2 equals impact angle.

position of the trajectory jet and the reverse roller associated with it could be reproduced in the flume. The flume was modified to vent the area below the nappe when desired and to measure the pressure in this region during unvented conditions. The unvented condition associated with accelerated erosion was reproduced by first increasing the discharge and then decreasing the discharge. The unvented condition was also established by raising the tailgate and flooding the knickpoint and then lowering the tailgate. The unvented highly turbulent condition was reproduced either by fluctuating the discharge or fluctuating the tail water elevation (May 1989). By venting and unventing the area under the nappe, the position of impact and the corresponding reverse roller can be positioned in a predictable and repeatable manner. The pressures measured in the area under the nappe ranged from .02 psi to .04 psi below atmospheric pressure to .03 psi above atmospheric pressure. The positive pressures were noted during periods of accelerated flow. These small negative pressures move the trajectory jet from a position where no significant headcutting can take place to a position very close to the vertical face of the knickpoint where the erosion potential is extremely high.

3.2 Geometric control

Calculations show that erosion is greater when the ratio of the height of fall, Z, to critical water depth, Y, is greater than 8 to 1 for a minimum tail water depth. As the ratio of Z/Y becomes larger the angle of impact (2) approaches 90° and a higher percent of the original discharge is entrained in the reverse roller portion of the jet (Q_3). The point of impact is controlled by the velocity of the flow coming over the knickpoint, for a given height of fall.

As the impact angle approaches 90° a higher percentage of the total discharge volume is entrained in and directed toward the vertical knickpoint face. The velocities of rock particles in the reverse roller were measured using a video recorder and stopwatch. The velocities averaged 0.71 ft/sec with a maximum velocity of 1.4 ft/sec.

Geology plays a critical role in determining the height of the overfall. In areas where the spillway or natural stream is underlain by "layered sedimentary units, the height of the overfall is controlled by the distance between one erosion resistant layer and the next underlying resistant layer. For example, in an area where the flow depth is typically 2 ft, a 16 ft section of resistant material overlying erodible material would yield a Z/Y ratio of 8/1. These conditions would bring the overfall near the vertical face.

Higher flow velocities move the point of impact further away from the vertical portion of the knickpoint and reduce the potential for undercutting. This is not intuitively apparent because in tractive force scour the higher the velocity the more severe the erosion. For the modeling effort it was decided to use the geometry and velocity that would bring the jet as close to the vertical face of the knickpoint as possible.

The mechanisms defined by this research offer an explanation as to why severe knickpoint migration occurs sporadically and at relatively low flows as opposed to at peak flows (Fig. 2). By evaluating the knickpoint as a vented or unvented drop structure, the conditions necessary to position the reverse roller for maximum undercutting can be predicted. Erosion thresholds are controlled by various combinations of knickpoint geometry and flow velocity. The geometry in turn is controlled by the site geology.

The point of impact of the water jet must be near the vertical face of the knickpoint for maximum undercutting. Under these conditions the angle, 2, at which the water jet strikes at the base of the knickpoint is approximately 90°. In a vented condition, when the ratio of Z/Y is 8/1 or greater and the tail water is low the angle of incidence approaches 90° and geometrically controlled erosion is enhanced.

In the unvented condition, the low-pressure zone below the nappe draws the water jet into the face of the knickpoint and maximizes the erosive effort of the reverse roller. These negative pressures below the nappe also have a significant impact on knickpoints developed in fractured competent rock units. Water is drawn through fractures or pores toward the low-pressure areas resulting in increased pore pressures and reduced shearing resistance.

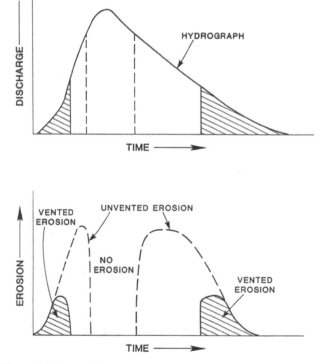

Figure 2. Unvented and vented knickpoint erosion thresholds for a hypothetical emergency/auxiliary spillway hydrograph.

3.3 Geological factors

Structural and stratigraphy discontinuities are critical elements in determining erosion potential. A discontinuity is an interruption in lithologic and physical properties in a rock mass (American Geological Institute 1972). Structural discontinuities are caused by natural compressive and tensional forces which alter rock mass properties. Discontinuities such as fractures, faults, joints, igneous dikes, and veins are common causes of knickpoint formation. Fractures are due to mechanical failure by stress and include cracks, joints, and faults. Faults are important features in causing erosion in general and knickpoint initiation in particular.

The majority of the erosion case histories studied for this research were associated with stratigraphic discontinuities. Stratigraphic discontinuities include stratified sedimentary rock sequences including those interbedded with volcanic and volcano-clastic rocks. These discontinuities include bedding planes, bed contacts, stratigraphic contacts, unconformities, pinchouts, facies changes, and sedimentary structures and textures (Cameron et al. 1986). Bedding planes are surfaces of deposition that visibly separate successive layers of stratified material (American Geological Institute 1972). Bedding planes often mark distinct boundaries between materials deposited in various environments of deposition. Coarse-grained basalt alluvial sand and gravel can have an abrupt contact with underlying shallow marine clay. Bedding planes play an important role in erosion, especially in the propagation of a head cut where a resistant layer overlies an erodible unit. Bedding planes are also important in conjunction with joints because at the intersections the strata are divided into blocks, which are susceptible to plucking during flood events.

Sedimentary structures are described by Berg (1986) as being the most characteristic features of sedimentary rocks and resulting from either primary stratification caused by sedimentary processes or secondary stratification produced by biological or physical changes shortly after deposition. Critically located sedimentary structures can initiate and influence the rate of erosion.

Unconformities are gaps in the geologic record that can bring rocks of vastly different compositions into abrupt contact. An unconformity which places a non-erodible material over an erodible one forms the same stratigraphic discontinuity as bedding contacts between materials of different erodibilities.

Pinch-outs and facies changes can cause abrupt lateral lithologic changes which are important factors in initiating a headcut. The rate of migration is often dependent on lateral facies changes and can increase or decrease dramatically as a result of these changes.

Litho-stratigraphic control is the "key" factor which controls the initiation and rate of erosion in stratified rock sequences. Sudden changes in stratigraphy, both vertically and laterally, control the geometry of the knickpoint, which in turn controls the severity of erosion.

The occurrence of a knickpoint is dependent on stratigraphic or other types of inhomogeneities in underlying materials. The location of the knickpoint is controlled by the occurrence of erosion.

The four basic types of mass failure observed and studied during the REMR research are:
– undercutting of a capping layer resulting in cantilever toppling of jointed or fractured rock,
– undercutting of caprock resulting in tensile failure and toppling of caprock,
– undercutting of erosion resistant layer resulting in shear failure of large blocks, and
– rafting of large blocks of jointed material as a result of water entering joints or fractures.

Knickpoint failure mechanisms described in this research are for two-layered geologic systems. One, or more, of these mechanisms probably occur during each flood event. The most dangerous erosive condition appears to occur when one dominant failure mechanism can act unimpeded along the spillway channel length.

Geologic factors are the least understood of the three major variables (hydrology, geometry, geology) and are independent of site hydraulics or channel geometry; often control spillway design; and, most significantly, govern all erosion processes (Mathewson 1998).

To provide analytical structure, a geologic erosion equation is proposed to concisely express critical geologic variables influencing erosion. The equation reads:

$$GR = f(L, SP, G, ST, MP) \tag{2}$$

where GR = geologic resistance to hydraulic erosion, L = lithology, SP = rock substance properties, G = genesis of the material, ST = structure and tectonic history, and MP = rock mass properties.

The following sections describe the role of each of the geologic factors and define an erodibility classification for each.

3.3.1 *Lithology*

Different erosion processes observed for various rock types indicate the first and most significant erosion criterion should be lithology. Lithology is the basic classification of natural material and relates to material genesis. The three general lithologic types are igneous, sedimentary, and metamorphic. Spillway flows have been documented during the REMR research on two of the three lithologic types, but by far the greatest frequency of flows has occurred on sedimentary rock. Clastic sedimentary rocks make up most observed sites. Relative to sedimentary rocks, igneous and metamorphic rocks tend to have higher strengths and densities and display competent mass properties. Igneous and metamorphic rocks display block-by-block detachment processes rather than single-grain erosion processes.

3.3.2 *Rock substance properties*

Rock substance properties involve properties of mineral grains and bonds between grains. These include density, strength, hardness, permeability, weathering, grain size, and grain shape. Observations indicate density, strength, and weathering play significant roles in the erosion process. These properties play a significant role in erosion of soils and very weak rock because these materials detach in a grain-by-grain process. With competent rock materials, the rock substance properties of cementation, strength, and density are better developed, and the mass properties become the controlling factors.

Greater amounts of weathering increase erodibility and changes density and strength. It is believed that a discussion of density and strength effects on erosion will indirectly address effects of weathering. Density influences entrainment by creating a particle too heavy to move. The erosion classification for rock substance is composed of both rock density and rock strength. These properties are related to the Unified Rock Classification System (URCS) and can be determined from this system.

3.3.3 *Genesis*

Processes that form or deposit geologic units determine the three-dimensional extent of each rock and soil bed. These processes are highly complex and produce materials which are anisotropic, heterogeneous, and discontinuous. The physical rock properties frequently change drastically along the length and/or width of the spillway channel. Knowledge of formational processes yields an understanding of material continuity and its properties. For example, sandstone, which formed from river sand, would be expected to be highly discontinuous, whereas marine shale should be more continuous.

The erosion classification based on genesis includes two components, vertical consistency and lateral consistency. Vertical consistency addresses the first-order discontinuities, thickness of each bed within each unit exposed in the channel. As bed thickness increases, the rock unit becomes more massive, and generally stronger, resulting in increased resistance and lower erosion potential. Lateral consistency is a measure of the total number of different rock facies or subunits exposed along the strike of a sedimentary bed. This factor addresses the problem of uniformity to erosion along the entire exposure of a unit. A uniform unit is expected to erode in such a manner that secondary knickpoints and other anomalies channel geometry will not develop and accelerate erosion.

3.3.4 *Structural and tectonic history*

The structural and tectonic history of an area controls the rock body orientation and the amount of fracturing in the rock unit. Orientation of the rock units can be horizontal or dipping. The most favorable orientations are horizontal or those that dip upstream toward the flow; units that dip downstream with the flow direction accentuate mass failures at the downstream end of the excavated channel and units that trend across the spillway channel and strike parallel to subparallel to the channel centerline cause highly complex erosion patterns resulting from channelization of the flow along bedding units which accentuates erosion. In cases where the spillway channel changes direction as it drains toward the valley, complex erosion patterns may also develop.

3.3.5 *Rock mass properties*

Rock mass properties are probably the most important of all in controlling erosion. Discontinuities such as fractures, joints, or bedding planes provide weaknesses along which detachment can take place. Postformational changes, second-order discontinuities, in the rock mass are generally termed rock mass properties. Another definition of rock mass properties is that they cannot be taken into the laboratory and measured because they are large-scale features, such as a fracture set or fold. Rock mass properties include bedding thickness, rock orientation and rock fracturing and can be mapped in an exploration trench or during excavation of the channel.

A direct result of the tectonic stresses placed on the rock mass is jointing and fracturing in a rock body. These discontinuities are weaknesses in an otherwise non-erosive surface. Hydraulic forces make use of these planes of weakness to pluck, slide, and otherwise detach rock particles from the mass. Massive unfractured rock is considerably more erosion resistant than is broken or shattered rock. Structural discontinuities are not random, but are related to the structural history of the site.

4 ROCK EROSION PREDICTION

Documentation of spillway flow events at numerous SCS, CE, and private dam provides a database from which to develop empirical relationships regarding earth material performance under hydraulic stresses. The erosion performance of an emergency/auxiliary spillway channel was found to be related to an interaction between the channel geometry, flow hydraulics, and site geology. Channel geometry and hydraulics are interrelated through the basic laws of hydrology. Site geology is an independent factor that often control channel geometry. Therefore, any technique to predict emergency/auxiliary spillway channel erosion potential must incorporate and evaluate all three primary factors in channel erosion. The following procedure is developed based on laboratory test and field observation mentioned in previous section for evaluating the integrity of unlined spillway. The process is performed by evaluating two indexes: Erosion Risk Class and Erosion Potential Class. The Erosion Risk Class is a rating based on the slope of spillway channel, peak flow velocity, and any geometric anomaly within the spillway channels. The Erosion Potential Class is a rating based on five geologic material behaviors that influence erosion resistance. The proposed procedure to estimate emergency/auxiliary spillway channel erosion potential is given below:

– Determine the Erosion Risk Classification (Fig. 3) for each segment of the channel based on the geometry and flow characteristics of the existing, proposed, or designed spillway channel. Incorporate any topographic anomalies, such as pilot channels, road fills, and gradient changes at the channel- valley boundary.

– Evaluate all site investigation data and field surveys. Using the Erosion Potential Classification (Fig. 4), determine the erosion potential class for each identifiable rock unit exposed or possibly exposed in the spillway channel for each of the factors in the geologic erosion equation (Lithology, Rock Substance Properties, Genesis, Tectonics, and Rock Mass Properties).

– Compare the Erosion RISK classification with the Erosion POTENTIAL Classification for each unit exposed within the channel. In each case where the POTENTIAL of a unit is less than the RISK (more erodible), special engineering geologic attention and design consideration are required for the unit. If the RISK is less than the POTENTIAL (More erosion resistant), the unit can be considered to be stable under the Proposed geometric and hydraulic conditions. It is important to recognize that this technique is empirical and that good engineering judgment will be required for each evaluation.

Figure 3 table:

EROSION RISK	EROSION RISK CLASS			
	AAAA	AAA	AA	A
Slope (percent)	30-45	15-30	4-15	< 4
Flow Velocity (ft/sec)	10-15	7-10	4-7	< 4
Geometric Anomaly	Extreme	Major	Moderate	None
AAAA	Significant Erosion Risk			
AAA	High Erosion Risk			
AA	Moderate Erosion Risk			
A	Slight Erosion Risk			

Figure 3. Erosion risk classification based on slope, flow velocity, and geometric anomaly within the spillway channel.

Figure 4 table:

EROSION POTENTIAL	EROSION POTENTIAL CLASS			
	AAAA	AAA	AA	A
AAAA	Erosion Resistant Rock			
AAA	Moderately Erosion Resistant Rock			
AA	Moderately Erodible Material			
A	Erodible Soil			
LITHOLOGY				
Sandstone	XXXX	XXXX	XXXX	
Shale & Siltstone		XXXXXX	XXXXXX	
Limestone	XXXXXXX			
Granular Soil (Low PI)				XXXXXX
Cohesive Soil (High PI)			XXXXXXXX	
Intrusive Igneous	XXXXXXX			
Extrusive Igneous	XXXXXXX	XXXXXXXX		
Massive Metamorphic	XXXXXXXXXX			
Foliated Metamorphic		XXXXXX	XXXXXX	
SUBSTANCE				
Density (pcf)	> 140	140-125	125-116	< 116
Uniaxial Strength (psi)	> 6000	6000-2000	2000-150	< 150
GENESIS				
Vertical Consistency (ft)	> 6	6-2	2-.25	< .25
Lateral Consistency (#)	1	2	> 2	> 2
TECTONICS				
Unit Orientation Related to Flow Direction	Flat	Dip Toward	Dip Parallel	Dip Away
ROCK MASS				
Fracture Spacing (ft)	> 3	3.0-1.0	1.0-0.5	< 0.5
Particle Diameter (ft)	3-5	1-3	1-.5	< .5
Fracture Size/Opening (in)	< 1/8	1/8-1/2	> 1/2	open/clean
Fracture Sets (No.)	2	2-3	> 3	shattered

Figure 4. Erosion potential classification based on lithology, rock substances, material genesis, postformational discontinuities (tectonics), and rock mass properties.

5 CONCLUSIONS

Based on the literature review, laboratory test results, and analyses of field data, the following conclusions can be drawn from this research:

-   The head cutting or knickpoint migration phenomenon is dependent on the geometry of the knickpoint and the velocity of the water. The geometry of the knickpoint is dependent on the geology at a specific site. The geometry of an undeveloped knickpoint can be predicted if enough geologic information is available. A rapidly moving knickpoint needs a relatively continuous resistant capping layer overlying a less resistant layer.
-   Techniques developed in this research can be used to hydraulically model certain aspects of headcut erosion using rock simulants. Rock simulants consisted of various combinations of sodium silicate cemented gravel, gelatin-cemented gravel, and plexiglas.
-   The highest rate of headcutting does not necessarily correspond to the highest velocity or discharge. Laboratory results have confirmed field reports that headcutting can be negligible at higher velocities and accelerate greatly as the velocity is reduced.
-   Unvented knickpoints in laboratory tests accelerated headcutting by orders of magnitude greater than vented knickpoints.
-   The diameter of the reverse roller portion of the jet remains relatively constant for the dimensions of the knickpoint and flow conditions. For the reverse roller to undercut effectively, the erodible layer has to be at least as thick as the diameter of the reverse roller.
-   The geologic resistance to erosion equation should apply to rock erosion at any sites impacted by high velocity jets of water.
-   The proposed spillway erosion prediction can be used as a quick evaluation on the integrity of unlined spillway channels.

REFERENCES

American Society of Civil Engineers/United States Committee On Large Dams. 1975. *Lessons from dam incidents, USA.* American Society of Civil Engineers, New York.

American Society of Civil Engineers/United States Committee On Large Dams. 1988. *Lessons from dam incidents, USA-11.* American Society of Civil Engineers, New York.

Annandale, G.W. 1995. "Erodibility." Journal of Hydraulics Research, Vol. 33, No. 4.

Berg, R. R. 1986. Reservoir Sandstones, Englewood Cliffs, NJ, Prentice-Hall, Inc.,

Blaisdell, F. W. 1983. Analysis of Scour Observations at Cantilever Outlets, Miscellaneous Publication No. 1427, US Department of Agriculture, Washington, DC.

Blaisdell, F. W., and Anderson, C. L. 1984. "Pipe Spillway Plunge Pool Design Equations," Proceedings, Conference on Water Resource Development, American Society of Civil Engineers, Hydraulics Division, pp 390-396.

Cameron, C. P., Cato, K. D., McAneny, C. C., and May, J. H. 1986. "Geotechnical Aspects of Rock Erosion in Emergency Spillway Channels," Technical Report REMR-GT-3, US Army Engineer Waterways Experiment Station, Vicksburg, MS.

Cameron, C. P., Patrick, D. M., Cato, K. D., and May, J. H. 1988a. "Geotechnical Aspects of Rock Erosion in Emergency Spillway Channels; Report 2: Analysis of Field and Laboratory Data," Technical Report REMR-GT-3, US Army Engineer Waterways Experiment Station, Vicksburg, MS.

Cameron, C. P., Patrick, M., Bartholomew, C. O., Hatheway, A. W., and May, J. H. 1988b. "Geotechnical Aspects of Rock Erosion in Emergency Spillway Channels; Report 3: Remediation," Technical Report REMR-GT-3, US Army Engineer Waterways Experiment Station, Vicksburg, MS.

Cato, K. D. 1991. "Performance of geological material under hydraulic stress," Ph.D. diss., Texas A&M University, College Station, TX.

Kirsten, H.A.D. 1982. "A Classification System of Excavation in Natural Materials." The Civil Engineer in South Africa, pp. 292-308, (discussion in Vol. 25, No. 5, May 1983).

Mason, P. J. 1984. "Erosion of Plunge Pools Downstream of Dam Due to the Action of Free-Trajectory Jets," Proceedings of the Institution of Civil Engineers, Paper 8734, Vol 76, pp 523-537.

Mathewson, C. C., Cato, K.D., and May, J. H. 1998. "Geotechnical aspects of rock erosion in emergency spillway channels, Supplemental Information on Prediction, Control, and Repair of erosion in Emergency Spillway Channels," Technical Report REMR-GT-3, US Army Engineer Waterways Experiment Station, Vicksburg, MS.

May, J. H. 1989. Geotechnical aspects of rock erosion in emergency spillways. Technical Report REMR-GT-3, Report 4 - Geologic and Hydrodynamic Controls on Headward Erosion, US Army Engineer Waterways Experiment Station, Vicksburg, MS.

Moore, J.S., Temple, D.M., and Kirsten, H.A.D. 1994. "Headcut Advance Threshold in Earth Spillways." Bulletin of the Association of Engineering Geologists, Vol. XXXI, No. 2, pp. 277-280.

Moore, W. L. 1943. "Energy Loss at the Base of a Free Overfall," Transactions of the American Society of Civil Engineers, Vol 108, p 1343, with discussions by White, M. P., p 1361, Rouse, H., p 1381, and others.

Reinius, E. 1986 (Jun). "Rock Erosion," Water Power and Dam Construction, pp 43-48.

Spurr, K. J. W. 1985 (Jul). "Energy Approach to Estimating Scour Downstream of a Large Dam," Water Power and Dam Construction, pp 81-89.

Temple, D. M. 1980. Tractive design of vegetated channels: Transactions of the ASAE, The American Society of Agricultural Engineers, St. Joseph, MI, Vol. 23, No. 4, pp. 884-890.

Temple, D. M. 1982. Flow retardance of submerged grass channel linings: Transactions of the ASAE, The American Society of Agricultural Engineers, St. Joseph, MI, Vol. 25, No. 5, pp. 1300-1303.

Temple, D. M. 1983. Design of grass-lined open spillways: Transactions of the ASAE, The American Society of Agricultural Engineers, St. Joseph, MI, Vol. 26, No. 4, pp. 1064-1069.

Temple, D. M. 1984. Erosionally Effective Soil Stress In Grass Lined Open Channels, The American Society of Agricultural Engineers Paper No. SWR84-102: The American Society of Agricultural Engineers, St. Joseph, MI, 13 p.

Temple, D. M. 1986. Distribution coefficients for grass-lined channels: Journal of Hydraulic Engineering, American Society of Civil Engineers, Paper No. 20435, Vol. 112, No. 3, pp. 193- 205.

Temple, D.M., and Hanson, G.J. 1993. AHeadcut Development in Vegetated Earth Spillways", 1993 International Summer Meeting of ASAE/CSAE, Spokane, Washington, 20-23 June, Paper No. 932017.

US Army Engineer District, Rock Island. 1962. "Spillway, Saylorville Reservoir, Design Memorandum No. 10," US Army Engineer District, Rock Island, Corps of Engineers, Rock Island, IA.

US Army Engineer District, Rock Island. 1984. "Saylorville Dam: Initial Overflow of Spillway, Supplement to Periodic Inspection Report No. 7," US Army Engineer District, Rock Island, Corps of Engineers, Rock Island, IA.

Vieux, B. E. 1986. "Plunge Pool Erosion in Cohesive Soils at Two Dams in Kansas," Draft report, US Department of Agriculture, Soil Conservation Service, Washington, DC.

Wibowo, J. L. and Murphy, W. 2002. "Unlined Spillway Erosion Prediction Model" ERDC Technical Report (in preparation).

Scour of rock due to high-velocity jet impact: a physically based scour model compared to Annandale's erodibility index method

E. Bollaert
Laboratory of Hydraulic Constructions, Swiss Federal Institute of Technology, Lausanne, Switzerland

G.W. Annandale
Engineering & Hydrosystems Inc, Highlands Ranch, Colorado, USA

A.J. Schleiss
Laboratory of Hydraulic Constructions, Swiss Federal Institute of Technology, Lausanne, Switzerland

ABSTRACT: A physically based engineering model has been developed at the Laboratory of Hydraulic Constructions for the evaluation of the ultimate scour depth of a jointed rock mass due to high-velocity jet impact. The model is based on experimental measurements and numerical simulations of water pressure fluctuations at plunge pool bottoms and inside underlying rock joints. The water pressures inside the joints revealed to be of highly transient nature, governed by a cyclic change between high peak pressures and low near-atmospheric pressures. Hence, the new engineering model, called the Comprehensive Fracture Mechanics (CFM) model, uses a simplified Linear Elastic Fracture Mechanics (LEFM) approach to express the erosion resistance of the rock mass. In the following, this model has been compared with Annandale's Erodibility Index (EI) method, which is actually a widely used semi-empirical model for prediction of the ultimate scour depth in jointed rock. A systematic comparison between the two methods allowed to express the sensibility of the different parameters of the new physically based model and points out the promising character of its results.

1 INTRODUCTION

The impact of high-velocity jets on jointed rock masses creates a progressive erosion of the rock by break-up of the joints and ejection of the so formed blocks. Prediction of this scour phenomenon has been a challenge for engineers since quite a long time. Most of the existing developments are based on model or prototype measurements of scour hole formation as a function of the hydraulic characteristics in question. Only a few methods consider the resistance of the rock mass against this erosion, most of them based on a sort of generalised description. One example is Annandale's Erodibility Index (EI) Method, which expresses the rock mass resistance by means of an index that accounts for the main geomechanical parameters. This method, developed in 1995 (Annandale, 1995; Annandale et al., 1998) has proven to be reliable and successful in many practical cases.

A new, completely physically based model has been developed at the Laboratory of Hydraulic Constructions of the Swiss Federal Institute of Technology for the prediction of the ultimate scour depth of jointed rock (Bollaert, 2002; Bollaert & Schleiss, 2002a; Bollaert & Schleiss, 2002b). This model, called the *Comprehensive Fracture Mechanics* (CFM) model, expresses the ultimate scour depth by a coupling of the predominant hydrodynamic and geomechanical parameters of the phenomenon. This is performed such that a practicing engineer can easily handle them, without neglecting basic physics behind it.

The aim of the present paper is to develop a sensibility analysis of the different parameters of this new model, based on the characteristics of Annandale's Erodibility Index (EI) Method. This points out the capabilities of the new model and its promising nature.

2 THE COMPREHENSIVE FRACTURE MECHANICS (CFM) MODEL

A new model for the evaluation of the evolution of the ultimate scour depth in intermittently or completely jointed rock, under the impact of high-velocity plunging jets, has been developed. The concept is based on the theoretical framework that has been outlined in Bollaert (2002) and Bollaert & Schleiss (2002a). As such, the model is completely physically based and represents a comprehensive assessment of the two major physical processes that govern break-up of rock: 1) hydrodynamic fracturing of rock joints, and 2) dynamic uplift of so formed rock blocks.

The model consists of three modules: 1) the falling jet, 2) the plunge pool, 3) the rock mass. Emphasis is given on the physical parameters that are necessary to accurately describe each phenomenon. The modules for the falling jet and for the plunge pool define the hydrodynamic loading that is exerted by the jet at the water-rock interface. The former determines the major characteristics of the jet from its point of issuance at the dam down to the point of impact into the plunge pool. The latter describes the diffusion of the jet through the plunge pool and defines the resulting jet excitation at the water-rock interface. The module for the rock mass has a two-fold objective. First of all, it transforms the hydrodynamic loading inside the joints into a critical stress intensity (for closed-end joints) or a net uplift impulsion (for single rock blocks). Four basic parameters describe the hydrodynamic loading inside closed-end or open-end rock joints:

1. maximum dynamic pressure coefficient C^{max}_p
2. characteristic amplitude of pressure cycles Δp_c
3. characteristic frequency of pressure cycles f_c
4. maximum dynamic impulsion C^{max}_I

The first parameter is relevant to brittle propagation of closed-end rock joints. The second and third parameters are necessary to express time-dependent propagation of closed-end rock joints. The fourth one is used to define dynamic uplift of rock blocks formed by open-end rock joints. All these parameters are extensively described in Bollaert (2002) and will not be further outlined herein. Secondly, the module of the rock mass defines the basic geomechanical characteristics, relevant for the determination of its resistance. Two failure criteria are relevant:

1. Failure of closed-end rock joints by *propagation* of the joints. This propagation can be instantaneous or time-dependent. The latter case involves failure by fatigue.
2. Failure of open-end rock joints by *dynamic uplift* or displacement of the rock blocks out of their surrounding mass.

Each of these criteria constitutes a physical limit for development of the scour hole. Which criterion is most restrictive is not evident and depends on the geomechanical characteristics of the rock mass. For practice, it is recommended to verify both of them, because they are strongly related one to the other. Rock blocks that cannot be ejected from their mass can still be subjected to break-up into smaller pieces (ball-milling). These smaller pieces could be entrained by the flow more easily. On the other hand, even when no further fracturing is possible, the rock mass might already been broken up and capable to be eroded by dynamic uplift. Hence, the CFM model only deals with the first failure criterion, i.e. progressive propagation of rock joints. In the following, it is assumed that this is the governing rock mass failure criterion.

Fracture Mechanics transforms the hydrodynamic loading into is a stress intensity factor K_I, and then compares this factor with the corresponding resistance of the material, which is expressed by the fracture toughness value K_{Ic}. The stress intensity factor is expressed as a function of the length of the crack a, the water pressure at the tip of the joint σ_{water}, and the nature and geometry of the surrounding rock, expressed by a boundary correction factor f:

$$K_I = \sigma_{water} \cdot f \cdot \sqrt{\pi a} \tag{1}$$

As described in Bollaert & Schleiss (2002), the boundary correction factor f is defined by the type of rock joint. Three main types have been distinguished: the semi-elliptical or semi-circular joint, the single-edge joint and the center-cracked joint (Figure 1).

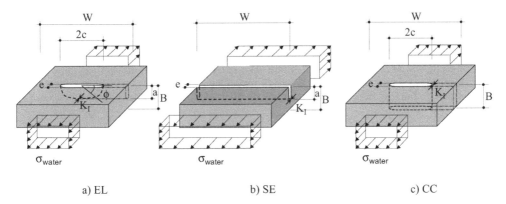

a) EL   b) SE   c) CC

Figure 1. Proposed framework for the basic geometrical configurations of intermittently jointed rock: a) semi-elliptical (EL) joint; b) single-edge (SE) joint; c) center-cracked (CC) joint.

The first crack is of semi-elliptical or semi-circular shape and, with regard to the applied pressure σ_{water}, partially sustained by the surrounding rock mass in two directions. As such, it is the geometry with the highest possible support of surrounding rock. Corresponding stress intensity factors should be used in case of low to moderately jointed rock. The second crack is single edge notched. Support from the surrounding rock is only exerted perpendicular to the plane of the notch and stress intensity factors will be substantially higher. Thus, it is more appropriate for moderately to highly jointed rock. The third geometry is center-cracked. Similar to the single edge notch, only one-sided rock support can be accounted for.

The *f* factor reflects the influence of the surrounding rock support and is presented in Figure 2 for the three geometries. At small a/B ratios, all configurations are quite close one to the other. At a/B ratios higher than 0.4, the elliptical notch (EL) stays at lower values, but these values are governed by the a/c and the c/W ratios. This is due to the two-sided character of the surrounding support. The main issue is that, with a careful choice of geometrical configuration, a first tendency of fracture propagation enhancement can already be distinguished.

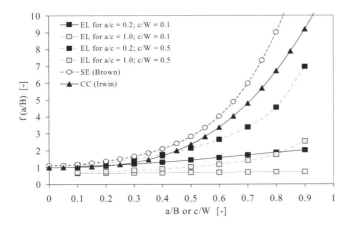

Figure 2. Different geometrical situations of joints on the theoretical stress intensity factor K_I.

Furthermore, the fracture toughness of the rock mass can be formulated as a function of the unconfined compressive strength (UCS) and the in-situ horizontal stress field σ_c as follows:

$$K_{I, UCS} = (0.010) \cdot UCS + (0.054 \cdot \sigma_c) + 0.42 \tag{2}$$

189

with UCS and σ_c in MPa and $K_{Ic,UCS}$ in MPa$\sqrt{m}$. Crack propagation distinguishes between brittle (or instantaneous) crack propagation and time-dependent crack propagation. The former happens for a stress intensity factor that is equal to or higher than the fracture toughness of the material. It occurs for both static and dynamic loadings and the governing velocity of cracking approaches the speed of sound in the medium. The latter is valid when the applied static or dynamic loading is inferior to the material's resistance. Cracks are propagated by stress corrosion or by fatigue. Failure by fatigue depends on the number and the amplitude of the load cycles.

3 ANNANDALE'S ERODIBILITY INDEX (EI) METHOD

Annandale's Erodibility Index (EI) Method compares the stream power of the water at the water-rock interface with the corresponding resistance of the rock against erosion.

The stream power of water is determined by the degree of jet energy dissipation through the plunge pool depth, which is expressed by the rate of velocity decay along the centerline of the jet. The rate of energy dissipation, or the stream power, is then expressed as a function of this velocity decay and can be compared with the stream power required to erode the rock (Figure 3b). The latter is defined by the "Erodibility Index" K_h, calculated based on the main geomechanical characteristics of the rock mass. The elevation where the available and required stream powers are equal is then considered as the ultimate scour depth.

The total rate of energy dissipation per unit area P [kW/m^2] is equal to the product of the unit weight of water γ [kN/m^3], the discharge Q [m^3/s] and the change in energy ΔE [m] divided by the horizontal projection of the area of the jet at impact A_i [m^2]:

$$P = \frac{\gamma \cdot Q \cdot \Delta E}{A_i} \tag{3}$$

The erodibility index K_h is a scalar number that is formed by the multiplication of four factors (Annandale, 1995). These factors account for the main mechanical characteristics of the rock mass, such as the unconfined compressive strength UCS, the relative density, the rock block size, the discontinuity bond shear strength, the shape of the rock blocks and the orientation of the joints relative to the impacting flow. The corresponding expression is written:

$$K_h = M_s \cdot K_b \cdot K_d \cdot J_s \tag{4}$$

in which:

M_s = the mass strength number
The mass strength number represents the material strength of an intact representative sample without regard to geologic heterogeneity within the mass. It equals the product of a material's uniaxial compressive strength (UCS) and its coefficient of relative density. The coefficient of relative density is the ratio of a material's bulk density over 27.0 kN/m^3. M_s can be determined by equating it to the UCS in MPa if the strength is greater than 10 MPa, and equal to $0.78 \cdot UCS^{1.5}$ when the strength is less than 10 MPa. As such, five different types of rock are distinguished, from very soft rock to extremely hard rock.

K_b = block size number
The block size number refers to the mean size of blocks of intact rock material (the cube root of the volume) as determined by the joint spacing within the rock mass. It can be calculated by dividing the RQD by the joint set number J_n. The RQD is the Rock Quality Designation, which is a standard parameter in drill core logging. The joint set number J_n is a function of the number of joint sets in a rock mass.

K_d = shear strength number
The shear strength number represents the strength of discontinuity interfaces in rock masses. It can be determined by the ratio J_r/J_a, in which J_r stands for the joint roughness number and J_a for the joint alteration number. The joint roughness number reflects the roughness condition of

the facing walls of a discontinuity. It is a parameter that depends on the tightness of the joints. The joint alteration number refers to the weathering condition of the joint face material.

J_s = relative ground structure number
The relative ground structure number accounts for the structure of the ground with respect to the direction of the incoming flow. In other words, it expresses the ease with which the flow can penetrate the ground and dislodge individual blocks, and is expressed in terms of the orientation and the shape of the individual blocks determined by joint set spacings, dip angles and dip directions.

All of the aforementioned parameters can be measured in the field at low cost and are quantifiable by means of tables. In the following, the Erodibility Index (EI) Method will be applied to different types of rock. For each type, the index will be compared with the erosive power of an incoming flow. This incoming flow is defined by its geometrical characteristics. As such, the pressure head of the flow that is necessary to just start scour can be determined for each of the rock masses in question. This pressure head, as well as the corresponding geometrical characteristics, is then used to compare with the present new scour model.

A log-log relationship between calculated rates of energy dissipation and the corresponding calculated erodibility index is presented in Figure 3a. The calculations are based on 150 field observations of scour and on published data on initiation of sediment motion. The correlation allows the prediction of a critical erosion threshold for any given set of hydraulic conditions and for any type of foundation material (granular soils, rock, etc.).

Recently, Annandale et al. (1998) have conducted an erosion experiment on near-prototype scale by means of a rectangular jet impinging on an artificially created fractured rock. The rock was simulated by two consecutive dipped layers of lightweight concrete blocks. The blocks were of rectangular shape and very flat. The obtained result confirmed the theoretically derived scour threshold.

The structure of this erosion study combines jet velocity decay both in the air and trough the plunge pool (empirically), jet and plunge pool aeration (empirically) and geomechanical characteristics of the rock mass (analytically). It perhaps actually constitutes the most pertinent and directly applicable evaluation method for the ultimate scour depth. However, despite its recent experimental validation, no direct dynamic parameters are incorporated and no physical background of rock break-up is evident in the model.

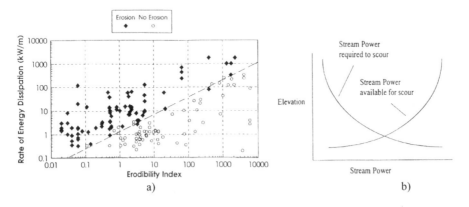

Figure 3. a) Rate of energy dissipation as a function of the erodibility index K_h (Annandale, 1995); b) Comparison of available and required stream power as a function of the location in the plunge pool.

4 COMPARISON BETWEEN THE CFM AND EI METHODS

The present example assumes a rock mass with one joint set that is vertically oriented and that is characterized by tightly healed joints. No plunge pool water depth is considered in the analysis. A vertically impinging two-dimensional jet of 2 m of width impacts at the rock-water interface, with a unitary discharge of q = 5.3 m²/s. Firstly, the necessary dam height (or jet fall height) necessary to scour the rock will be outlined based on the EI Method. This is done as a function of the strength of the rock mass. Secondly, for a chosen strength of the rock mass, a sensibility analysis of the CFM Model is developed.

4.1 Annandale's Erodibility Index Method

A rock mass with one joint set corresponds to a J_n value of 1.22 (Annandale, 1995). The joint is of the closed-end type. The distance between the joints is such that they do not interfere with each other. Hence, a very high RQD value of 95 is assumed. This results in a K_b number of 78. Furthermore, tightly healed joints are considered, with a discontinuous pattern, leading to a J_r number of 4 and a J_a number of 0.75 to 1.0. The dip direction of the joints is chosen in the direction of the falling jet. The dip angle is 90°, corresponding to J_s numbers ranging form 1.14 to 1.26, depending on the ratio of joint spacing. The value 1.14 is representative for equi-sided rock blocks, while the value of 1.26 rather represents rectangular strips.

The jet thickness at impact is assumed equal to 2 m, for a unitary discharge of q = 5.3 [m²/s]. From this, the pressure head H [m] that is necessary to start the scour is:

$$H = \frac{P \cdot 2}{\gamma \cdot 5.3} \tag{5}$$

P stands for the required stream power per unit area (eq. (3)). It can be noticed from Table 1 that, for very soft rock, a pressure head of only 10 m is sufficient to scour the rock. For extremely hard rock, the necessary heads increase up to 200 m.

Rock type		Very Soft		Soft		Hard		Very Hard		Extremely Hard	
M_s	[-]	3.3		13		26		106		280	
RQD	[-]	95		95		95		95		95	
J_n	[-]	1.22		1.22		1.22		1.22		1.22	
K_b	[-]	78		78		78		78		78	
J_r	[-]	4		4		4		4		4	
J_a	[-]	0.75	1	0.75	1	1	1.05	1.1	1.15	1.2	1.25
K_d	[-]	5.3	4.0	5.3	4.0	4.0	3.8	3.6	3.5	3.3	3.2
J_s	[-]	1.26	1.14	1.26	1.14	1.14	1.116	1.092	1.068	1.044	1.02
Erodibility Index K_h	[-]	1727	1172	6803	4616	9232	8607	32776	30662	75875	71166
Power Required P	[kW/m²]	268	200	749	560	942	894	2436	2317	4572	4357
Necessary Head H	[m]	10	8	29	22	37	35	94	90	177	169

Table 1. Erodibility Index K_h for different types of rock and stream power P required to erode the rock. Comparison with the pressure head H that is necessary to generate the required stream power.

4.2 Comprehensive Fracture Mechanics Model

The next step is to apply the Comprehensive Fracture Mechanics model to the same situation. This has been done at Table 2 for different degrees of jointing. The considered rock mass is assumed to have an unconfined compressive strength of 106 MPa, corresponding to very hard rock at Table 1. Semi-elliptical and semi-circular surface cracks are assumed. The a/c ratio is taken equal to 0.2 and 1.0. The total joint lengths L are 1.0 m or 2.0 m and the maximum dynamic pressure value at the tip of the joint C^{max}_p is defined as 1.0, 1.5 or 2.0 times the kinetic energy of the impacting jet.

Where instantaneous or brittle crack propagation has been obtained this has been indicated by "I". For a time-dependent crack propagation of up to 200-300 days of discharge, the results are indicated in light grey. This case is considered to represent where, during the lifetime of the dam structure, the rock mass will be scoured at some point in time. In other words, it is a broad-brush expression for the ultimate scour depth or, in the present case, for the pressure head or dam height that will cause scour. Finally, where time-dependent crack propagation is obtained with more than 200-300 days of discharge, the results are indicated in dark grey. The latter case is considered here to represent the end of scour.

Some very interesting tendencies and sensitivity of the scour model can be distinguished. First of all, the maximum dynamic pressure coefficient C^{max}_p has a profound impact on the scour limit. Secondly, the degree of jointing also significantly influences the results. As such, a C^{max}_p of 1.0 combined with a degree of jointing of 0.2 and an a/c ratio of 0.2 (elliptical flaw) will result in a necessary dam height of 120 m. The same C^{max}_p but with a degree of jointing of 0.6 will commence with scour of the rock at a lower dam height of 80 m.

It can be observed that these results do not hold in the case of a circular joint. Apparently, the semi-circular joint is much more difficult to erode than the semi-elliptical one. This corresponds to the findings in Figure 2 on boundary correction factors f.

P_e	f(a/W)	a/c	H (EI)	H (CFM)	C_{max} = 1.0 L = 1.0 m T	C_{max} = 1.5 L = 1.0 m T	C_{max} = 2.0 L = 1.0 m T	C_{max} = 1.0 L = 2.0 m T	C_{max} = 1.5 L = 2.0 m T	C_{max} = 2.0 L = 2.0 m T
-	-	-	m	m	days	days	days	days	days	days
		0.2		80	14400	116	3.7	933	7.5	I
		0.2		90	3550	29	0.9	230	1.8	I
0.2	1.116	0.2	90-94	100	1000	8	0.26	65	I	I
		0.2		110	326	2.6	I	21	I	I
		0.2		120	116	0.9	I	7	I	I
		0.2		80	60	0.5	I	3.9	I	I
		0.2		90	15	I	I	I	I	I
0.4	1.294	0.2	90-94	100	4.2	I	I	I	I	I
		0.2		110	1.3	I	I	I	I	I
		0.2		120	0.4	I	I	I	I	I
		0.2		80	0.6	I	I	I	I	I
		0.2		90	I	I	I	I	I	I
0.6	1.642	0.2	90-94	100	I	I	I	I	I	I
		0.2		110	I	I	I	I	I	I
		0.2		120	I	I	I	I	I	I
		1.0		80	6500000	52000	1700	421000	3360	110
		1.0		90	1600000	12800	419	104000	828	27
0.2	0.668	1.0	90-94	100	457000	3650	120	29600	236	7.7
		1.0		110	147000	1170	38	9520	76	2.5
		1.0		120	52000	417	13.7	3380	27	I
		1.0		80	136000	1090	36	8800	71	2.3
		1.0		90	33600	269	8.8	2170	17	I
0.4	0.676	1.0	90-94	100	9580	77	2.5	620	5	I
		1.0		110	3080	25	0.8	199	1.6	I
		1.0		120	1090	9	I	71	I	I
		1.0		80	7400	60	2	5350	3.8	I
		1.0		90	1800	15	0.5	1320	I	I
0.6	1.107	1.0	90-94	100	521	4.2	I	376	I	I
		1.0		110	168	1.3	I	121	I	I
		1.0		120	59	0.5	I	43	I	I

Table 2 Sensitivity analysis of the main parameters of the Comprehensive Fracture Mechanics model by comparison with Annandale's Erodibility Index Method. The analysis makes use of semi-elliptical and semi-circular shaped single joint. The UCS of the rock mass is 106 MPa.

The interesting aspect about Table 2 is that a comparison with Annandale's Erodibility Index Method procures different combinations of parameters that are in agreement with the former. For example, for a total joint length L of 1.0 m and a semi-elliptical shape (a/c = 0.2), a plausible combination would be a persistency a/B of 0.4 and a C^{max}_p of 1.0. This results in 10 to 15 days of discharge before the scour starts, which is a realistic value. Another possibility is obtained for a total joint length L of 1.0 m and a semi-circular shape (a/c = 1.0), which results in 5 to 9 days of discharge for a persistency a/B of 0.4 and a C^{max}_p of 2.0. For a total joint length L of 2.0 m and a semi-circular shape (a/c = 1.0), a persistency a/B of 0.4 and a C^{max}_p of 1.5 conduct to 10 to 17 days of discharge to scour the rock mass.

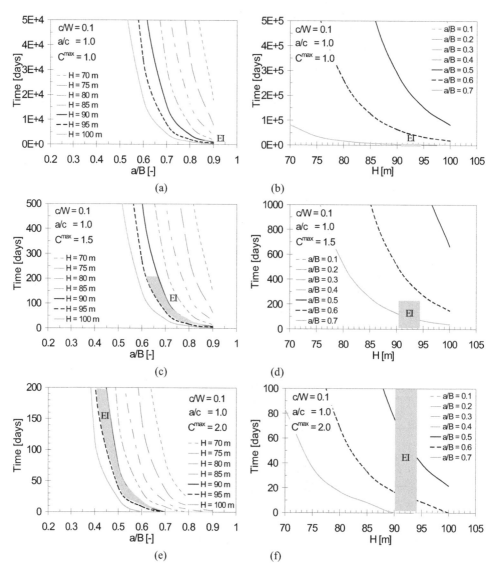

Figure 4. Time necessary to scour a very hard rock mass (UCS = 106 MPa) with a single semi-circular joint:

    a) as a function of a/B with C^{max} = 1.0;   b) as a function of H with C^{max} = 1.0;
    c) as a function of a/B with C^{max} = 1.5;   d) as a function of H with C^{max} = 1.5;
    e) as a function of a/B with C^{max} = 2.0;   f) as a function of H with C^{max} = 2.0.

The sensibility of the a/B ratio, the a/c ratio, the c/W ratio and the C^{max} coefficient has been studied more in detail. The results are presented in Figure 4 and consider the case of a *semicircular* joint, i.e. a/c = 1.0. Furthermore, according to the singular joint assumption made in Annandale's Erodibility Index Method, the c/W ratio has to be taken very small. A value of 0.1 has been chosen, which corresponds to a longitudinal distance W between the successive rock joints of four times the joint width 2c.

The left hand side of Figure 4 deals with the time necessary to completely break-up the joint as a function of the a/B ratio, i.e. the persistency of the rock mass. This has been done for three C^{max} coefficients: 1.0, 1.5 and 2.0. For the fatigue law, 80 % of this maximum pressure has been used. At each graph, the curves corresponding to pressure heads between 70 m and 100 m are compared. Assuming core jets and, thus, no significant water cushion, these heads can be considered equal to the fall height of the jet.

According to Table 2, a very hard rock mass with a UCS of 106 MPa and a RQD of 95 has been used. The Erodibility Index for this rock indicates that the pressure head H that is necessary to scour the rock lies *between 90 and 94 m*. It is assumed that this scour is obtained within maximum 200 days of discharge on the rock. These limits have been systematically added to the graphs for purpose of comparison.

When considering the curves for 90 m and 95 m of pressure head in Figure 4a, it can be derived that a C^{max} coefficient of 1.0 needs a degree of jointing of 0.9 to scour the rock within 200 days of discharge. A C^{max} coefficient of 2.0 (Figure 4e) would need an initial degree of jointing of only 0.4 to 0.7, which seems much more plausible. A similar conclusion can be drawn from the Figures 4b & 4f, as a function of the necessary pressure head H.

The issue is to determine which is a plausible maximum pressure coefficient. This coefficient is defined as the sum of the C-coefficient (mean pressure coefficient) and the C^+-coefficient (extreme positive deviation from the mean pressure). Maximum C^+ coefficients generally indicate a value of about 2.0 at a Y/D_j equal to 0 (= without any water cushion) (Bollaert, 2002). The corresponding C-coefficient of mean pressure is equal to 0.85. Hence, an upper limit for the C_{max} coefficient should be around 3. The corresponding results are presented hereunder.

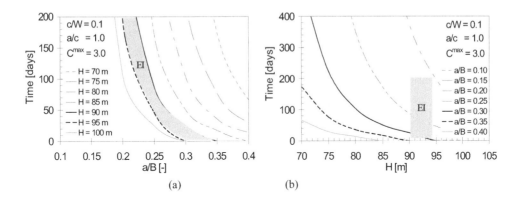

Figure 5.   Time necessary to scour a very hard rock mass (UCS = 106 MPa) with a single semi-circular joint: a) as a function of a/B with C^{max} = 3.0; b) as a function of H with C^{max} = 3.0.

It is concluded from Figure 5 that, in order to obtain a pressure head that is similar to the Erodibility Index method, the present scour model combines a maximum pressure coefficient C^{max} of 3.0 with an initial degree of jointing of 0.25 to 0.35. This result is valid for a single semi-circular rock joint inside a very hard rock mass.

A similar sensibility analysis has been performed for a single *semi-elliptical* rock joint. The a/c ratio has been taken equal to 0.5. This means that the vertical depth of the rock joint is equal to half of the horizontal radius. All other parameters are the same as in the previous analysis.

The results are presented in Figure 6 and mainly indicate that the semi-elliptical rock joint is slightly more sensible to crack propagation than the semi-circular one. A C^{max} coefficient of 1.0 (Figure 6a) needs a degree of jointing of only 0.65 to scour the rock within 200 days of discharge. A C^{max} coefficient of 2.0 (Figure 6e) would need an initial degree of jointing of only 0.3 to 0.4.

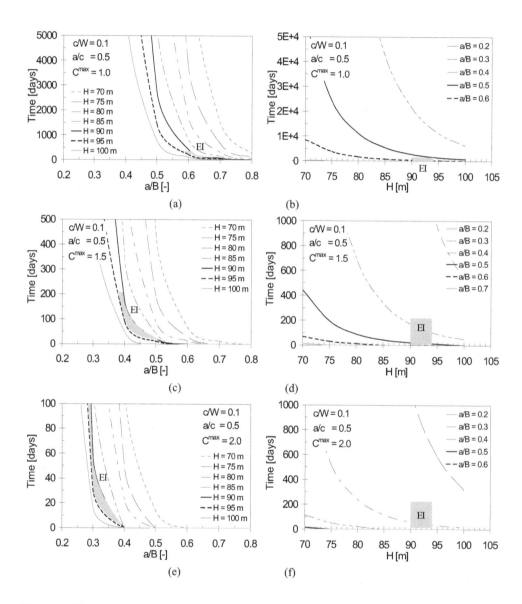

Figure 6. Time necessary to scour a very hard rock mass (UCS =106 MPa) with a single semi-elliptical joint:

a) as a function of a/B with C^{max} = 1.0;   b) as a function of H with C^{max} = 1.0;
c) as a function of a/B with C^{max} = 1.5;   d) as a function of H with C^{max} = 1.5;
e) as a function of a/B with C^{max} = 2.0;   f) as a function of H with C^{max} = 2.0.

Finally, the case of a *single-edge* rock joint has been investigated. As already pointed out before, this type of joint has much less support from the surrounding rock mass and, therefore, should be much easier to break-up. This is confirmed by the results in Figure 7, where a C^{max}-coefficient of 2.0 corresponds to a degree of jointing of only 0.15 to 0.20. These values are significantly lower than for the other two types of rock joints. When introducing the critical degree of jointing that was obtained for the semi-elliptical joint, i.e. 0.3 to 0.4, brittle crack propagation is obtained, even at pressure heads of only 70 m.

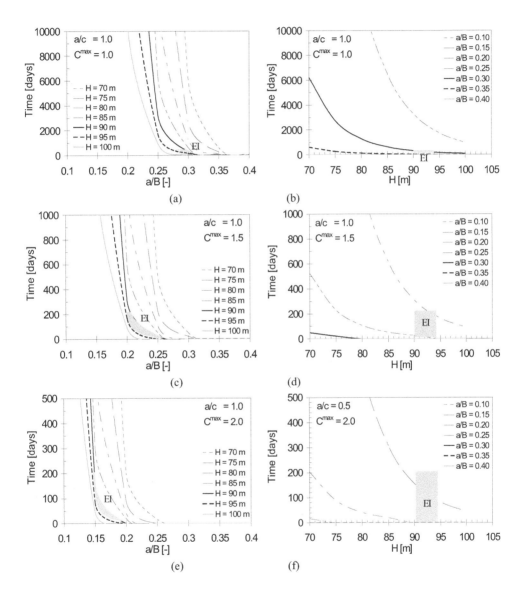

Figure 7.   Time necessary to scour a very hard rock mass (UCS = 106 MPa) with a single single-edge joint:
a) as a function of a/B with C^{max} = 1.0;   b) as a function of H with C^{max} = 1.0;
c) as a function of a/B with C^{max} = 1.5;   d) as a function of H with C^{max} = 1.5;
e) as a function of a/B with C^{max} = 2.0;   f) as a function of H with C^{max} = 2.0.

The conclusions drawn from the preceding calculations are only valid for a very hard rock with a UCS = 106 MPa. A similar analysis is performed for a hard rock and a *semi-circular* joint. According to Table 2, this rock has a UCS = 26 MPa. The Erodibility Index for this rock indicates that the pressure head H that is necessary to scour the rock lies *between 35 and 37 m*. The results are presented in Figure 8. The relationship between the C^{max} coefficient and the appropriate degree of jointing is in good agreement with the corresponding relations found for the very hard rock mass.

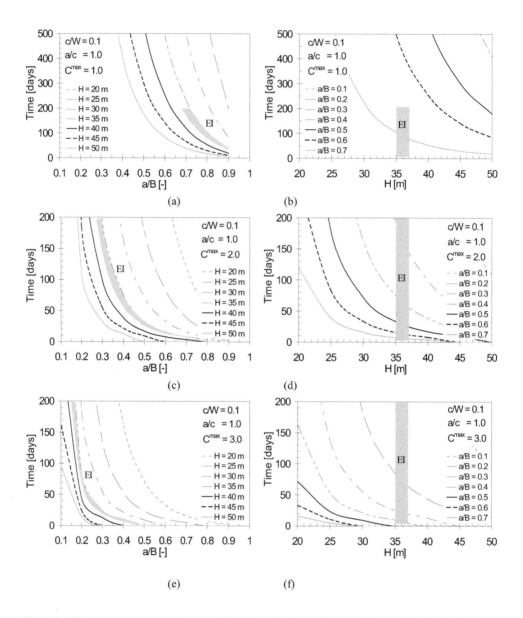

Figure 8.   Time necessary to scour a hard rock mass (UCS = 26 MPa) with a single semi-circular joint:
a) as a function of a/B with C^{max} = 1.0;   b) as a function of H with C^{max} = 1.0;
c) as a function of a/B with C^{max} = 2.0;   d) as a function of H with C^{max} = 2.0;
e) as a function of a/B with C^{max} = 3.0;   f) as a function of H with C^{max} = 3.0.

5 CONCLUSIONS

The present comparison between the Comprehensive Fracture Mechanics (CFM) Model and Annandale's Erodibility Index (EI) Method indicate that an interesting agreement can be found between the two methods as a function of the different parameters of the CFM model. The agreement gives a first idea about the sensibility of these parameters and can be used as basis to perform case studies on real dams.

6 REFERENCES

Bollaert, E. 2001. Spectral density modulation of plunge pool bottom pressures inside rock fissures. *Proceedings of the XXIXth IAHR Congress*, Student Paper Competition, Beijing.

Bollaert, E. & Schleiss, A. 2001a. A new approach for better assessment of rock scouring due to high velocity jets at dam spillways. *Proceedings of the 5th ICOLD European Symposium*, Geiranger, Norway.

Bollaert, E., Pirotton, M., Schleiss, A. 2001c. Multiphase transient flow and pressures in rock joints due to high-velocity jet impact: an experimental and numerical approach. *Proceedings of the 3rd International Symposium on Environmental Hydraulics*, Tempe, 3-8 December 2001.

Ewing, D.J.F. 1980. Allowing for Free Air in Waterhammer Analysis. *Proceedings of the 3rd International Conference on Pressure Surges*, Canterbury, England.

Martin, C.S. & Padhmanaban, M. 1979. Pressure Pulse Propagation in Two-Component Slug Flow. Transactions of the ASME, *Journal of Fluids Engineering*, Vol. 101, pp. 44-52.

Whiteman, K.J. & Pearsall, I. S. 1959. Reflux Valve and Surge Tests at Kingston Pumping Station, *British Hydromech. Res. Assoc., National Engineering Laboratory joint report* N°1.

Wylie, E.B. & Streeter, V.L. 1978. Fluid transients, *McGraw-Hill Inc.*, US.

Rock Scour due to falling High-velocity Jets - Schleiss & Bollaert (eds)
© 2002 Swets & Zeitlinger, Lisse, ISBN 90 5809 518 5

Quantification of the relative ability of rock to resist scour

G.W. Annandale
Engineering & Hydrosystems Inc., Highlands Ranch, Colorado, USA

ABSTRACT: The paper summarizes a methodology that can be used to quantify the relative ability of rock to resist scour. Rock's relative ability to resist scour is a function of its intact material strength, block size, shear strength of discontinuities between blocks of rock, and the relative shape and orientation of the rock blocks. A geomechanical index that is known as the Erodibility Index is used to quantify the relative ability of rock to resist scour, and is calculated as the product of four coefficients, each representing one of the four parameters. The orientation and shape of rock blocks are incorporated in one coefficient.

A summary of the Erodibility Index Method that implements this measure of the relative ability of rock to resist scour is presented in a companion paper to this volume (Annandale 2002a) as is the approach to quantify scour extent (Annandale 2002c).

1 INTRODUCTION

The Erodibility Index Method (Annandale 1995) relates the relative ability of earth material to resist scour, expressed in terms of a geomechanical index known as the Erodibility Index, to the relative magnitude of the erosive power of water, expressed in terms of stream power, to define an erosion threshold for an extensive range of earth materials. The procedure that is used to calculate the Erodibility Index for rock is outlined in this paper. Procedures for quantifying this index for other earth materials can be found in Annandale (1995).

2 THE ERODIBILITY INDEX

The Erodibility Index (K) that is used to quantify the relative ability of earth material to resist scour is identical to Kirsten's Excavatability Index (Kirsten 1982) and is expressed as:

$$K = M_s \cdot K_b \cdot K_d \cdot J_s \qquad (1)$$

The intact mass strength number (M_s) represents the strength of a homogenous, "perfect" sample of earth material. In order to acknowledge the roles of discontinuities and imperfections for determining the earth material's relative ability to resist scour, the intact mass strength number is multiplied by other parameters. The value of the intact mass strength number is adjusted by multiplying it with the block / particle size number (K_b), the discontinuity / inter-particle bond shear strength number (K_d) and the relative ground structure number (J_s). Ways to quantify each of these numbers are presented in what follows.

3 INTACT MASS STRENGTH NUMBER FOR ROCK

The values of M_s for rock are related to conventional descriptions of rock hardness, field identification tests and the unconfined compressive strength (*UCS*) of the rock (Table 1). The latter can be quantified by making use of procedures described in ASTM D-2938 (Standard Test Method for Unconfined Compressive Strength of Rock Core Specimens).

Table 1. Mass strength number for rock (M_s)

Hardness	Identification in Profile	Unconfined Compressive Strength (MPa)	Mass Strength Number (M_s)
Very soft rock	Material crumbles under firm (moderate) blows with sharp end of geological pick and can be peeled off with a knife; is too hard to cut tri-axial sample by hand.	Less than 1.7	0.87
		1.7 – 3.3	1.86
Soft rock	Can just be scraped and peeled with a knife; indentations 1 mm to 3-mm show in the specimen with firm (moderate) blows of the pick point.	3.3 – 6.6	3.95
		6.6 – 13.2	8.39
Hard rock	Cannot be scraped or peeled with a knife; hand-held specimen can be broken with hammer end of geological pick with a single firm (moderate) blow.	13.2 – 26.4	17.70
Very hard rock	Hand-held specimen breaks with hammer end of pick under more than one blow.	26.4 – 53.0	35.0
		53.00 – 106.0	70.0
Extremely hard rock	Specimen requires many blows with geological pick to break through intact material.	Larger than 212.0	280.0

The values of M_s for rock can also be quantified by making use of the equations listed below.

$$M_s = C_r \cdot (0.78) \cdot (UCS)^{1.05} \text{ when } UCS \le 10 \text{ MPa} \tag{2}$$

and

$$M_s = C_r \cdot (UCS) \text{ when } UCS > 10 \text{ MPa} \tag{3}$$

where C_r = coefficient of relative density. In the case of rock $C_r = g.\rho_r / (27.10^3)$ with ρ_r = mass density of the rock in kg/m^3; and $g = 9.82$ m/s^2, the acceleration due to gravity; 27.10^3 N/m^3 = unit weight of good quality rock.

202

4 BLOCK SIZE NUMBER FOR ROCK

The block size number for rock (K_b) acknowledges the roles of the sizes of rock block in determining earth material resistance to scour. Increases in rock block size offers increased resistance to scour.

Joint spacing and the number of joint sets within a rock mass determines the value of K_b. Joint spacing is estimated from borehole data by means of the rock quality designation (RQD) and the number of joint sets is represented by the joint set number (J_n). RQD is a standard parameter in drill core logging and is determined as the ratio between the sum of the lengths of rock pieces longer than 0.1 m and the total core run length (usually 1.5 m), expressed as a percent (Deere and Deere 1988). RQD values range between 5 and 100. A RQD of 5 represents very poor quality rock, and a RQD of 100 represents very good quality rock. For example, if a core contains four pieces of rock longer than 0.1m, with lengths of 0.11 m, 0.15 m, 0.2 m and 0.18 m then the cumulative length of rock longer than 0.1 m is 0.64 m and the RQD is 0.64 m / 1.5 m x 100 = 43.

Schematic presentations explaining the joint set concept are shown in Figure 1 and in the photographs in Figures 2 and 3. The values of the joint set numbers (J_n) are found in Table 2. J_n is a function of the number of joint sets, ranging from rock with no or few joints (essentially intact rock), to rock formations consisting of one to more than four joint sets. The classification accounts for rock that displays random discontinuities in addition to regular joint sets. Random joint discontinuities are discontinuities that do not form regular patterns. For example, rock with two joint sets and random discontinuities is classified as having two joint sets plus random (see Table 2). Having determined the values of RQD and J_n, K_b is calculated as

$$K_b = \frac{RQD}{J_n} \qquad (4)$$

where $5 \leq RQD \leq 100$ and $1 \leq J_n \leq 5$.

With the values of RQD ranging between 5 and 100, and those of J_n ranging between 1 and 5, the value of K_b ranges between 1 and 100 for rock.

If RQD data is unavailable, its value can be estimated with one or more of the following equations:

$$RQD = (115 - 3.3 \cdot J_c) \qquad (5)$$

J_c is known as the joint count number, a factor representing the number of joints per m^3 of the material, calculated as,

$$J_c = \left(\frac{3}{D}\right) + 3 \qquad (6)$$

where D = mean block diameter in m.

D can be calculated with the equation:

$$D = (J_x \cdot J_y \cdot J_z)^{0.33} \text{ for } D \geq 0.10 \text{ m} \qquad (7)$$

Where J_x, J_y and J_z = average spacing of joint sets in m measured in three mutually perpendicular directions, x, y and z. Joint set spacing can be determined by the Fixed Line Survey (see e.g. International Society for Rock Mechanics 1981, Geological Society of London 1977, Bell 1992).

Other equations used to calculate RQD, derived from those above, are,

$$RQD = \left(105 - \frac{10}{D}\right) \qquad (8)$$

and

$$RQD = \left(105 - \frac{10}{\left(J_x \cdot J_y \cdot J_z\right)^{0.33}}\right) \qquad (9)$$

Table 2. Joint set number (J_n).

Number of Joint Sets	Join Set Number (J_n)
Intact, no or few joints/fissures	1.00
One joint/fissure set	1.22
One joint/fissure set plus random	1.50
Two joint/fissure sets	1.83
Two joint/fissure sets plus random	2.24
Three joint/fissure sets	2.73
Three joint/fissure sets plus random	3.34
Four joint/fissure sets	4.09
Multiple joint/fissure sets	5.00

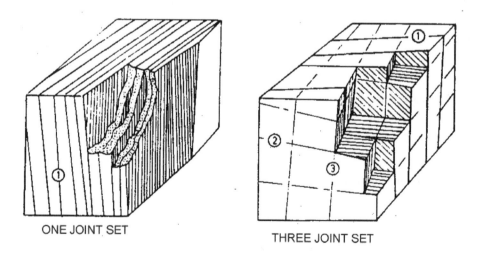

ONE JOINT SET

THREE JOINT SET

Figure 1. Schematic presentations illustrating the concept of joint sets.

Figure 2. A rock formation with one joint set

Figure 3. A rock formation with three joint sets

205

5 DISCONTINUITY BOND SHEAR STRENGTH NUMBER

The discontinuity bond shear strength number (K_d) is the coefficient that represents the relative strength of discontinuities in rock determined as the ratio between joint wall roughness (J_r) and joint wall alteration (J_a):

$$K_d = \frac{J_r}{J_a} \qquad\qquad (10)$$

J_r represents the degree of roughness of opposing faces of a rock discontinuity, and J_a represents the degree of alteration of the materials that form the faces of the discontinuity. Alteration relates to amendments of the rock surfaces, for example weathering or the presence of cohesive material between the opposing faces of a joint. Values of J_r and J_a can be found in tables 3 and 4. The values of K_d calculated with the information in these tables change in sympathy with the relative degree of resistance offered by the joints. Increases in resistance are characterized by increases in the value of K_d. The shear strength of a discontinuity is directly proportional to the degree of roughness of opposing joint faces and inversely proportional to the degree of alteration.

Table 3. Joint roughness number (J_r)

Joint Separation	Condition of Joint	Joint Roughness Number
Joints/fissures tight or closing during excavation	Stepped joints/fissures	4.0
	Rough or irregular, undulating	3.0
	Smooth undulating	2.0
	Slickensided undulating	1.5
	Rough or irregular, planar	1.5
	Smooth planar	1.0
	Slickensided planar	0.5
Joints/fissures open and remain open during excavation	Joints/fissures either open or containing relatively soft gouge of sufficient thickness to prevent joint/fissure wall contact upon excavation.	1.0
	Shattered or micro-shattered clays	1.0

Table 4. Joint alteration number (J_a)

Description of Gouge	Joint Alteration Number (J_a) for Joint Separation (mm)		
	1.0^1	$1.0-5.0^2$	5.0^3
Tightly healed, hard, non-softening impermeable filling	0.75	-	-
Unaltered joint walls, surface staining only	1.0	-	-
Slightly altered, non-softening, non-cohesive rock mineral or crushed rock filling	2.0	2.0	4.0
Non-softening, slightly clayey non-cohesive filling	3.0	6.0	10.0
Non-softening, strongly over-consolidated clay mineral filling, with or without crushed rock	3.0	6.0**	10.0
Softening or low friction clay mineral coatings and small quantities of swelling clays	4.0	8.0	13.0
Softening moderately over-consolidated clay mineral filling, with or without crushed rock	4.0	8.00**	13.0
Shattered or micro-shattered (swelling) clay gouge, with or without crushed rock	5.0	10.0**	18.0
Note: [1]Joint walls effectively in contact. [2]Joint walls come into contact after approximately 100-mm shear. [3]Joint walls do not come into contact at all upon shear. **Also applies when crushed rock occurs in clay gouge without rock wall contact.			

Joint roughness is described by referring to both large and small-scale characteristics. The large-scale features are known as stepped, undulating or planar; whereas the small-scale features are referred to as rough, smooth or slickensided. Examples of planar and undulating joints are shown in Figure 4 and Figure 5 respectively. Figure 6 is a schematic presentation of conventional descriptions of joint roughness.

A planar, rough joint indicates that the large-scale feature is planar, but that the joint surfaces are rough. The concept of closed, open and filled joints, terminology used in table 3, is illustrated in Figure 4. The value of K_d that is calculated by means of equation (10) is roughly equal to the tangent of the residual angle of friction between the rock surfaces.

Figure 4. Planar joints.

Figure 5. Undulating joints.

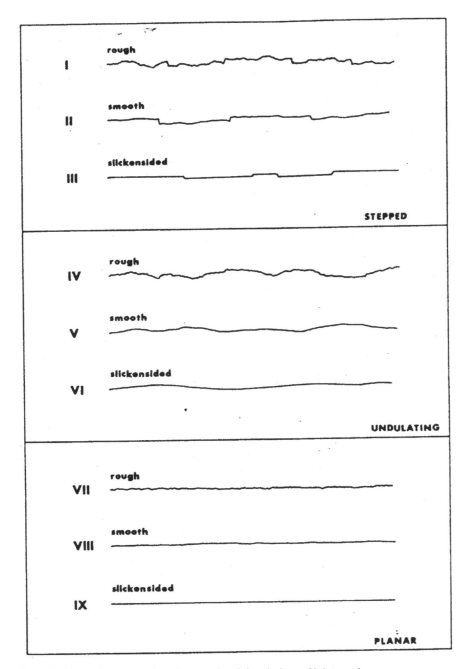

Figure 6. Schematic presentation of conventional descriptions of joint roughness.

6 RELATIVE GROUND STRUCTURE NUMBER

The relative ground structure number (J_s) represents the relative ability of earth material to resist scour due to the structure of the ground. This parameter is a function of the dip and dip direction of the least favorable discontinuity (most easily eroded) in the rock with respect to the direction of flow, and the shape of the material units. These two variables (orientation and shape) affect the ease by which the stream can penetrate the ground and dislodge individual material units.

The concepts of dip and dip direction of rock are illustrated in Figure 7. This figure shows a perspective view of a block of rock with a slanting discontinuity. The line that is formed where the horizontal plane and the plane of the discontinuity intersect is known as the strike of the rock. The dip direction, measured in degrees azimuth, is the direction of a line in the horizontal plane that is perpendicular to the strike and located in the vertical plane of the dip of the rock. The dip of the rock is the magnitude of the angle between the horizontal plane and the plane of the discontinuity, measured perpendicular to the strike. If the flow direction is roughly in the same direction as the dip direction, then the dip is said to be in the direction of the flow. If the flow direction is opposite to the dip direction, then the dip is said to be opposite to the direction of flow.

The shape of rock blocks is characterized by determining the joint spacing ratio (r), which is the quotient of the average spacing of the two most dominant high angle joint sets in the vertical plane – see Figure 8.

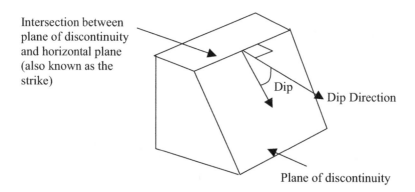

Intersection between plane of discontinuity and horizontal plane (also known as the strike)

Dip

Dip Direction

Plane of discontinuity

Figure 7. Definition sketch pertaining to dip and dip direction of rock.

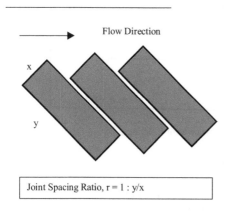

Flow Direction

x

y

Joint Spacing Ratio, r = 1 : y/x

Figure 8. Determination of the joint spacing ratio, *r*.

210

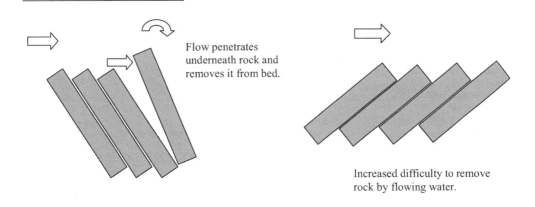

Flow penetrates
underneath rock and
removes it from bed.

Increased difficulty to remove
rock by flowing water.

Figure 9. Influence of dip direction on scour resistance offered by rock.

Conceptually the function of Relative Ground Structure Number (J_s), incorporating shape and orientation, is as follows. If rock is dipped against the direction flow, it will be more difficult to scour the rock than when it is dipped in the direction of flow. When it is dipped in the direction of flow, it is easier for the flow to lift the rock, by penetrating underneath and removing it. Rock that is dipped against the direction of flow will be more difficult to dislodge (Figure 9).

The shape of the rock, represented by the ratio r, impacts the erodibility of rock in the following manner. Elongated rock will be more difficult to remove than equi-sided blocks of rock (figure 10). Therefore, large ratios of r represent rock that is more difficult to remove because it represents elongated rock shapes.

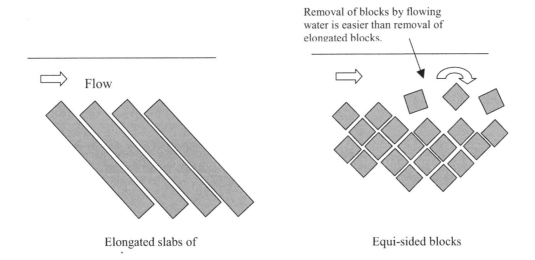

Removal of blocks by flowing
water is easier than removal of
elongated blocks.

Flow

Elongated slabs of

Equi-sided blocks

Figure 10. Influence of shape of rock blocks on scour resistance.

Once the effective dip, the direction of the effective dip relative to the flow direction (with or against), and the joint spacing ratio have been determined, use Table 5 to determine J_s. When working with intact material, such as massive rock, the value of J_s is 1.0. In cases where the value of r is greater than 8, use the values of J_s for $r = 8$.

Table 5. Relative ground structure number (J_s)

Dip Direction of Closer Spaced Joint Set (degrees)	Dip Angle of Closer Spaced Joint Set (degrees)	Ratio of Joint Spacing, r			
		1:1	1:2	1:4	1:8
180/0	90	1.14	1.20	1.24	1.26
In direction of stream flow	89	0.78	0.71	0.65	0.61
	85	0.73	0.66	0.61	0.57
	80	0.67	0.60	0.55	0.52
	70	0.56	0.50	0.46	0.43
	60	0.50	0.46	0.42	0.40
	50	0.49	0.46	0.43	0.41
	40	0.53	0.49	0.46	0.45
	30	0.63	0.59	0.55	0.53
	20	0.84	0.77	0.71	0.67
	10	1.25	1.10	0.98	0.90
	5	1.39	1.23	1.09	1.01
	1	1.50	1.33	1.19	1.10
0/180	0	1.14	1.09	1.05	1.02
Against direction of stream flow	-1	0.78	0.85	0.90	0.94
	-5	0.73	0.79	0.84	0.88
	-10	0.67	0.72	0.78	0.81
	-20	0.56	0.62	0.66	0.69
	-30	0.50	0.55	0.58	0.60
	-40	0.49	0.52	0.55	0.57
	-50	0.53	0.56	0.59	0.61
	-60	0.63	0.68	0.71	0.73
	-70	0.84	0.91	0.97	1.01
	-80	1.26	1.41	1.53	1.61
	-85	1.39	1.55	1.69	1.77
	-89	1.50	1.68	1.82	1.91
180/0	-90	1.14	1.20	1.24	1.26
Notes:	1. For intact material take J_s = 1.0				
	2. For values of r greater than 8 take J_s as for r = 8				

7 REFERENCES

Annandale, G.W. 2002a. The Erodibility Index Method: An Overview, *International Workshop on Rock Scour*, EPFL, Lausanne, Switzerland, September 25-28, 2002.
Annandale, G.W. 2002c. Quantification of Extent of Scour using the Erodibility Index Method, *International Workshop on Rock Scour*, EPFL, Lausanne, Switzerland, September 25-28, 2002.
Annandale, G.W. 1995. Erodibility, *Journal of Hydraulic Research*, Vol. 33, No. 4, pp. 471-494.
Kirsten, H.A.D., 1982, A classification system for excavation in natural materials, *The Civil Engineer in South Africa*, July, pp. 292-308.

Rock Scour due to falling High-velocity Jets - Schleiss & Bollaert (eds)
© 2002 Swets & Zeitlinger, Lisse, ISBN 90 5809 518 5

Wave impact induced erosion of rock cliffs

G. Wolters
Research Assistant, The Queen's University of Belfast, Civil Engineering Department, Belfast, UK

G. Müller
Lecturer, The Queen's University of Belfast, Civil Engineering Department, Belfast, UK

ABSTRACT: Rock cliffs and blockwork coastal structures often suffer wave induced damages. A peculiar aspect of these damages is that the damages involve the seaward removal of individual blocks from blockwork structures, or the breaking off of large blocks of rock along fissure or crack lines inside the rock, leading to undercutting. The assumption that wave impact pressures travel into water filled fissures, thus generating the damages, was only recently confirmed at Queen's University Belfast (QUB). An experimental study showed that wave impact pressures travel as compression waves into water filled cracks, whereby the pressure pulses rapidly decreased in magnitude. Real cracks in rock cliffs are however often only partially filled with water. A new experimental study, also conducted at QUB, revealed that wave impact generated pressures can travel into fully or partially water filled cracks or joints, thus damaging or destroying the structure from within. In partially submerged cracks the pressure pulse was found to travel in the air, propagating fast and with little attenuation deep into the structure. This finding implies that partially filled cracks are probably more dangerous than completely filled cracks. The propagating pressure pulses may therefore be the main cause for the removal of blockwork in coastal engineering structures or of rock cliff material.

1 INTRODUCTION

Many coastal structures such as breakwaters or rock cliffs contain cracks or fissures which extend above and below the mean water line, and which are exposed to wave attack. During storms, the effect of wave action on the cracks and the 'structure' becomes much more violent when waves start to break against the crack entrance, generating high but short pressure peaks of 10-100 ms duration. The effect of storms on rock cliffs and the houses on top of the cliff are described e.g. in Benumof and Griggs (1999): *"Waves ... were extremely powerful, often "shaking" and "rattling" the cliffs."*. Later it is stated that *"Cliffs with many open joints, where water can compress air and cause recoil are more subject to erosion than those which are relatively free from such openings."* In the context of damages to rocks and blockwork coastal structures, Shields (1895) states: *'Wherever joints occur, either in rock or in artificial structures, both mechanical and chemical action proceeds the fastest. Apart from the inherent weakness of joints, the air or water confined within them, when struck by a wave, is converted into a very destructive agent'*. These facts were repeated again, in a more modern language, in a more recent textbook. *'One practical indication of shock forces against concrete sea walls is the manner in which ill-designed or badly constructed lift joints become rapidly exploited by the sea'*, Muir Wood & Fleming (1981).

Cliff erosion can have many causes: increased ground water level, salt weathering, watertable fluctuations, biogenic activity, piezometric pressure changes, freeze-thaw action, just to name a few. The action of waves is however mostly held not directly responsible or even important for cliff failure: Carter states that there are numerous studies of coastal cliff morphology that never mention waves at all (Carter, 1991).

In engineering the damages caused by wave action on coastal structures, dams, ships, off-shore platforms are however well known. The action of breaking waves against such structures is hereby considered to constitute the major cause of damages. In particular older coastal structures, built from blockwork, contain cracks or joints which are exploited during storms. Typical damage mechanisms of blockwork structures and rock cliffs include the *seaward* removal of individual blocks in breakwaters or, in the case of rock cliffs, the breaking off of large blocks around mean water level, undercutting the cliff. This type of damages lead to the suspicion that wave impact induced pressures are acting not only from the outside, but also in the inside of the structure or rock cliff.

Although in the last 20 years a lot of research effort was directed towards the investigation of wave impact pressures, very little was known about impact pressure *propagation* and the pulse *characteristics*. Müller (1997) was the first to demonstrate that these impact pressures can actually enter water filled cracks or fissures and that they have the characteristics of a compression wave. Following his work, further studies were carried out to analyze pressure pulse propagation in completely submerged cracks, e.g. Müller et al. (2002). In a very different context, namely the erosion of rock underneath plunge pools, it was recently shown by Suisse researchers that compression waves entering water filled cracks in rocks can erode the rock severely, Bollaert & Schleiss (2001). In rock cliffs however, completely filled cracks are not typical; crack networks with both partially and fully submerged cracks are found more often. Such partially filled cracks are the topic of this study.

2 REVIEW

The authors' field of research originally centered on blockwork breakwaters with rubble filling which were built during the 19th century. Fig. 1a shows a typical cross section, and Fig. 1b a view (Le Havre breakwater) of such structures. Damages are often caused by the seaward removal of individual blocks during storms, with a subsequent loss of integrity of the blockwork and removal of further blocks. Although blockwork breakwaters are not representative for all coastal structures they share common features with rock cliffs and other old engineering structures: they all contain joints and are often composed of brittle materials with high compressive but low tensile strength. All are exposed to breaking wave action. The implication is that similar damage mechanisms occur in all of them.

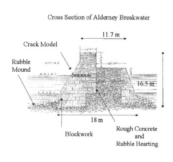

a. Typical cross – section

b. Le Havre Breakwater

Fig 1: Cross section and side view of blockwork breakwaters

2.1 Erosion by wave action?

One of the reasons why wave action has not attracted much attention in geologic circles is illustrated in the following statement made by Carter (see also Wilson,1952, Paskoff 1978, Paskoff & Sanlaville 1978, references in Carter 1991): *Many rock types are immensely strong, so that wave forces may have little effect. Many cliffed coastlines have a primary tectonic control through jointing and faulting and are little altered by wave action.*"

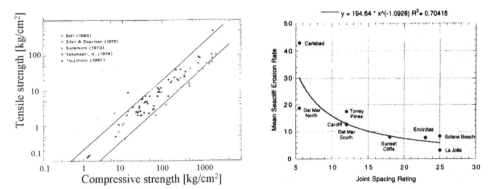

a: Tensile rock strength, Sunamura (1992)   b: Erosion vs. joint spacing, Benumof & Griggs (1999)

Fig. 2: Rock strength and erosion patterns

Fig. 2a shows the tensile rock strength plotted against the compressive strength. It can be seen that the compressive strength of rock is generally 10 times higher than the tensile strength and that the tensile strength (varying between 0.01-10 MPa) is however well in the range of water impact pressures (1MPa, see e.g. Rouville 1938). The actual pressure of a breaking wave acting against a rock face can thus be expected not to do any damage, except when stones are hurled by the wave against the rock. This damage mechanism would however be expected to leave mostly ground down rock - sand, and not large blocks (as observed on the coastline) as a residue.

Fig. 2b shows sea cliff erosion rates at various sites in California. The graph shows the cliff erosion rate plotted against joint spacing. It can be seen that erosion rates increase for increasing joint spacing, indicating some relationship between these two parameters. The pictures in Fig. 3 were taken at the Normandy coastline in France, near Le Havre (Etretat). Fig. 3a shows chalk cliffs exposed to marine erosion leaving in its wake arches and spikes (cliff retreat), and Fig. 3b a cliff undercutting. The cliffs are about 100m high and composed of chalk layers, some with flint flint inlay, which are sometimes separated by darker and much harder layers of dolomite chalk. Between the chalk layers wide cracks can be observed, Fig. 3b, which extend a couple of meters into the cliff, showing that the layer boundaries are very susceptible to wave attack. Fig. 3c, taken at St. Pierre en Port, shows a house on top of a cliff which faces imminent cliff collapse. In the foreground a recent rock fall can be observed.

a. Marine erosion   b. Cliff undercutting

c.: Cliff collapse

Fig 3: Cliff erosion

Breaker types and wave impact pressures

In the literature, four different types of breakers are defined as indicated in Fig. 4. Of these four breakers, only the plunging breaker can generate the high breaking wave impact loadings so detrimental to coastal structures. The spilling breaker, typical for very shallow sea bed slopes (tan α < 1:50, where α is the angle of the sea bed with the horizontal), curls over long before reaching the structure, so that only a highly aerated mass of turbulent water which does not generate a 'hard' impact ever reaches the structure.

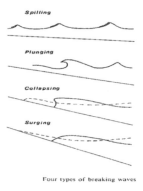

Fig. 4: Breaker types

The collapsing and the surging breaker (tan α > 1:3), occurring on very steep slopes, does not really 'break' at all; the wave collapses in itself thus dissipating its energy. The plunging breaker (1:5 < tan α < 1:50) has the typical shape associated with a breaking wave, with a jet of water emerging from the wave crest and hitting the coastal structure or rock cliff, generating very high impact pressures. These pressures are characterised by very short but high pressure pulses which are followed by a significantly smaller hydrodynamic pressure. Rise times of 0.005 sec and pressure magnitudes of up to 690 kPa were reported by Rouville (1938). Recent measurements on Alderney breakwater recorded pressures of up to 435 kPa, Bullock et al. (1999).

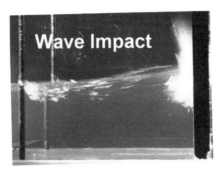

Fig. 5: Breaking wave impact (Model tests at QUB)

Fig. 5 shows a typical wave impact at model scale (breaking wave height approximately 60mm). The picture clearly shows the violence of the impact of the breaker tongue. Currently, it is very difficult to predict pressure magnitudes; formulae available in text books are unreliable at best. This is caused by the fact that the mechanism of pressure generation during impact is still not known precisely. A block which was separated from the structure or rock cliff can very often be easily moved by the waves. During the breakwater failure at Sines, Portugal in 1978 concrete blocks protecting the sea wall, each weighing 42t, were displaced and moved by wave action.

Only six field measurements of wave impact pressures are reported in the literature. Rouville (1938) recorded the highest impact pressures to date at Dieppe (France). The highest pulses recorded were very short, with a rise time of 5ms, and a magnitude of 690 kPa. From the facts reported so far, the following conclusions can be drawn:
(i)   Breaking waves generate impact pressure pulses of high magnitude and short duration.
(ii)   The erosion rate of rock cliffs depends on joint width and spacing.
(iii)   Wave impact pressures can travel as compression pulses into water filled cracks.
(iv)   Blockwork coastal structures as well as cliffs appear to be susceptible to damages created by wave impact pressures entering water filled cracks.
In addition to water filled cracks, fissure systems in rock can be expected to be only partially filled with water, so that not only the propagation of pressure pulses through water but also through air and water is of interest.

217

3 MODEL TESTS AT QUB

3.1 *Aims of model tests*

Model tests were conducted to address the following main objectives:
(i)   Can pressures propagate into partially water filled cracks?
(ii)   What are the characteristics of the pulse?
(iii)   Can blocks be removed by impact induced internal pressures?

3.2 *Experimental set-up*

A series of experiments was conducted in the Hydraulics Laboratory at the Queen's University of Belfast's (QUB) Civil Engineering Department in a wave tank of 17m length, 350 mm width and with a water depth of 1m. An inserted false bottom made of fibreglass brought the water depth from 1m at the deep end to 110 mm at the model sea wall with a slope of 1:10. The tank is shown in Fig. 7. At the shallow end of the sea bed, a vertical wall was installed. Waves were generated with a flap-type wave paddle in the deep water section of the tank. A single wave, with a deep water wave height of 71mm, was generated every 82 seconds. This procedure allowed to generate repeatable impact pressures within acceptable limits.

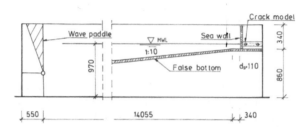

Fig. 7: Wave tank (dimensions in mm)

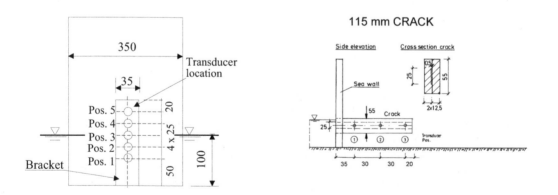

a.   Front wall with transducer bracket   b.   115mm crack, side view

Fig. 8: Sea wall model, all dimensions in mm.

Two brackets were manufactured which were inserted into the vertical seawall alternatively. The first bracket with fiove transducer positions (No. 5 was not instrumented since the waves

218

did not generate any pressures at that elevation) was used to record the wave impact pressures on the seawall (see Fig. 8a). The second bracket allowed any of the crack configurations to be securely inserted into the wall. The crack widths varied between 0.5, 1 and 3mm and the height was constant at 25mm. Crack lengths of 115 and 600 mm were employed. Cracks were examined while being totally submerged as well as partially submerged. Submergence varied between 0.1 h, 0.5 h, and 0.9 h (h = crack height). Fig. 8b shows the positioning of the bracket in the model wall. All the apparatus was made of stiff Perspex.

4 EXPERIMENTAL RESULTS

4.1 *Wave impact pressures*

Initially, the impact pressures created by the breaking wave on the vertical wall were recorded. Fig. 9 shows a typical impact pressure record for the transducer positions 1-4 as indicated in Fig. 8a; Fig. 10 shows the peak pressures for two series of 15 measurements.

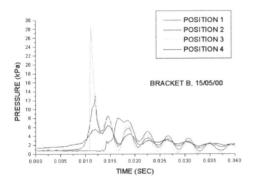
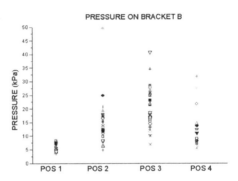

Fig. 9: Pressures on front wall   Fig. 10: Vertical distribution of impact pressures

From Fig.'s 9 and 10 it can be seen that the impact pressures are of comparatively short duration and that – despite of the single breaker technique – a large variability of peak pressures exists. The pressure magnitude ranges from approximately 5 kPa to 40 kPa.

4.2 *Pressure propagation*

115mm crack

In the fully submerged 115mm crack, the pressure traces in Fig. 11a show that the pulse travels at a speed of around 100 m/s. Positions 1 and 2 (entrance and centre of crack) show a double-peak signal, indicating that the pressure pulse was reflected at the end of crack, subsequently traveling out again. Position 3 shows a higher pressure peak than that at positions 2 and 3 because of the superposition of the incoming and the reflected pulse. Fig. 11a also indicates that the pressure pulse attenuates rapidly.

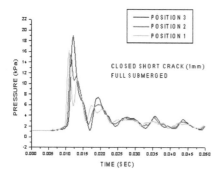 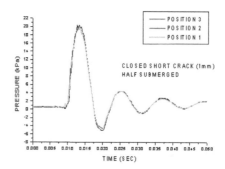

a.: P-T-Trace of fully submerged crack b.: P-T-Trace of half submerged crack

Fig. 11: Experimental Results for fully and partially submerged 115mm crack

600mm crack

Another series of experiments was conducted with longer cracks in order to assess the propagation characteristics of the wave impact generated pressure pulses. Fig. 12 shows a typical result from a 600mm long crack of 0.5 × 10mm cross-section. The speed of propagation in this case could be determined as 65 m/s. It can be seen that the pressure pulse attenuates fast. The pulse loses magnitude and increases in duration, indicating that significant energy dissipation is taking place. The low propagation velocity could only be explained by the presence of small air bubbles inside of the water which change the compressibility (and thus the speed of sound) in the water dramatically. The trace at x = 550mm indicates a reflection of the pressure pulse at the closed end of the crack. In addition, it was found that the high frequency content of the pressure pulse was damped out preferentially. The water-air mixture contained in the crack thus constitutes a 2-phase medium with very different properties when compared with pure water. Other experiments with larger cracks of up to 18.5 × 10mm showed maximum speeds of propagation of 200 – 250 m/s, see e.g. Müller et al. (2002).

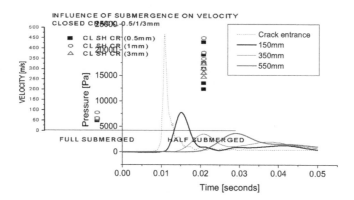

Fig. 12: Pressure pulse propagation through 600mm crack

220

Partially submerged 115mm crack

In a new series of experiments, the crack model was located with its center line at mean water level, so that the crack was only partially filled with water. In Fig. 11b, practically no time lapse between the different transducer positions can be seen. From the pressure record, the speed of propagation could not be properly determined since the data acquisition rate of 10 kHz was not sufficient to show enough detail; a value of 300 m/s could be estimated. Fig. 11b therefore appears to show the elastic response of the air enclosed in the crack. These observations gave rise to the following conclusions:

(i)   The pressure pulse seems to constitute an elastic wave *travelling in the air* rather than in the water at the speed of sound.

(ii)   Partially filled cracks transport wave impact pressures *fast and deep* into the inside of the structure and behind the protective blockwork.

(iii)   The high propagation velocity and the low attenuation within partially filled cracks indicates that partially filled cracks may be more dangerous than completely water filled ones.

Fig. 13 shows the comparison of the propagation velocities for partially and completely submerged cracks. The data acquisition rate of 10 kHz did not give sufficient data points to evaluate velocities above 300 m/s accurately, resulting in some spread of values around this value.

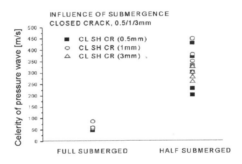

Fig. 13: Speed of propagation for fully and partially submerged crack

Originally it was thought that pressures propagate fast through water, being the denser medium, and slowly through air, but the experiments showed otherwise: low velocities were found for the full submerged crack whereas high velocities were found for the half submerged case. So it is actually the air, which behaves stiffer than the water, transporting the pressures faster and deeper into the structure.

5 EROSION MECHANISM OF FISSURED ROCK

In rock cliffs, it can be visualized that crack growth is caused by a process called wedge action (Fig. 14). When struck by a wave, the pressure inside the crack builds up rapidly, forcing an opening of the crack and focusing the tensile stresses at the crack tip, thus causing the failure of the brittle rock to crack there. This leads to a progressive growth of the crack. The propagation of pressures into cracks therefore leads to an attack onto the material where it is weakest, in tension. The brittleness of the material means that even short pressure pulses may lead to a failure in tension at the crack tip. Loose blocks can then be easily removed from the cliff face by wave action, a process often referred to as quarrying or plucking.

Wedge Action

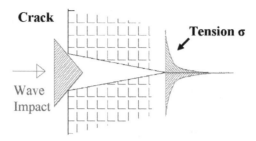

Fig. 14: Wedge Action

The rock coast shown in Fig. 4a shows a zig-zag line in plan view, indicating that the failure of the coastline occurs along the main fissure lines rather than being continuously eroded, supporting the assumption of a wedge-type erosion mechanism.

6 DISCUSSION

A review of the engineering literature showed that the exposure risk of a coastal structure to breaking wave action is a function of wave steepness and sea bed slope; only plunging breakers generate the high impact pressures held responsible for damages to such structures. Model tests indicated that wave impact pressures can travel into partially water filled cracks with the speed of sound in air and without any discernible damping, generating seaward pressures in block-work structures and splitting forces in fissured rock. This mechanism can be assumed to contribute significantly to damages to coastal structures as well as to the undercutting of rock cliffs. Due to the high compressive strength of rock when compared with recorded wave impact pressures, direct wave action against rock faces appears not to constitute a significant erosive force. One peculiar aspect of wave impact driven undercutting seems to be that – since breaking waves only occur in storm events – the erosion (and subsequent failure) of rock cliffs occurs not continuously but possibly only within a few days in every year. Further research in this field is required in order to assess exposure condition and erosion rate, to investigate the propagation of impact pressures into water or air filled crack systems and to relate rock undercutting to these influences.

7 CONCLUSIONS

Wave action of rock cliffs is not usually considered as one of the main erosive forces on rock cliffs. The erosion mechanism usually assumed is wave pressure induced abrasion of the rock front. Within a study of wave induced damages of blockwork coastal structures it was found that wave impact generated pressures can travel into water filled cracks or joints, thus damaging or destroying the structure from within. A similar damage mechanism may apply to rock cliffs, generating splitting pressures inside of the rock thus straining the material where it is weakest, in tension. In rock cliffs, cracks are however continuous systems and may well be only partially filled with water. A series of experiments was conducted in QUB's wave channel to assess the characteristics of breaking wave impact induced pressure pulse propagation through fully and partially water filled cracks and thus to determine the possibility of this erosion mechanism for rock cliffs. In completely submerged cracks the pressure pulses generally travel at very low velocities of 70-100 m/s. This slow speed of propagation for completely submerged cracks was attributed to the fact that the water constitutes a two-phase-medium with very different properties than pure water (air content of approximately 1.0%). Velocities in partially submerged cracks were found to be around 300 m/s; significantly higher than in water filled cracks and in the range of the speed of sound in air. It was also found that pulses attenuate fast inside the fully submerged crack, and slowly in partially submerged ones. The impact of waves on partially filled cracks or crack systems would thus allow the propagation of high and short pressure pulses deep into the system with little attenuation of the pulse. Pressures applied within a crack will enforce an opening of the crack and lead to very high tensile stresses at the crack tip, causing the brittle rock material to fail at this location and leading to crack growth. It is hypothesized that partially water filled cracks which are exposed to breaking waves are possibly even more dangerous for the integrity of blockwork structures and rock cliffs than water filled cracks.

ACKNOWLEDGEMENTS

The authors would like to gratefully acknowledge the financial support from the German Academic Exchange Board (*Deutscher Akademischer Austauschdienst*, DAAD) and from the UK's Engineering and Physical Science Research Council EPSRC under Grant GR/ M49755.

REFERENCES

Benumof, B.T. & Griggs, G.B.,1999, *The Dependence of Seacliff Erosion Rates on Cliff Material Properties and Physical Processes: San Diego County, California*, Shore & Beach, 67, No.4, pp. 29-41.

Bollaert E. & Schleiss A., 2001, A new approach for better assessment of rock scouring due to high velocity jets at dam spillways, *Proc. 5th ICOLD Symp.*, Geiranger / Norway.

Bullock G.N., Hewson P., Crawford A.R., Bird P.A.D., 1999, Field and laboratory measurements of wave loads on vertical breakwaters, *Proc. Coastal Structures 99*, Santander/Spain.

Carter R.W.G., 1991, *Coastal Environments: An Introduction to the Physical, Ecological and Cultural Systems of Coastlines*, Academic Press, London, 617p.

Chadwick A., MorfettJ., 1999, *Hydraulics in Civil and environmental engineering*, E & FN Spon, London, 3rd Ed.

Muir Wood A.M., Fleming C.A., 1981, *Coastal Hydraulics*, 2nd edition, The MacMillan Press Ltd., London, p 216.

Müller G., 1997, Propagation of wave impact pressures into water filled cracks, *Inst.* Civ. Engnr's, Water and Maritime, Vol. 124, Issue 2, 79-85.

Müller G., Cooker M. J., Allsop W., Bruce T., Franco L., 2002, Wave effects on blockwork structures model tests, *Journal of Hydraulic Research*, **40**, 117-124

Müller G., Cooker M. & Wolters G., 2002, Pressure-pulse propagation through aerated water in a crack (manuscript, submitted to Proc. Roy. Soc. A).

Rouville, M.A., 1938, Etudes internationales des efforts dus aux Lames, *Annales des Ponts et Chaussees*, **108**, 5-113.

Shield W., 1895, *Principles and Practice of Harbour Construction*, Longmans, Green & Co., London.

Sunamura, T. 1992, *Geomorphology of Rocky Coasts*, John Wiley & Sons, Chichester, 302 p.

Numerical modeling of rock scour

Physical model study of Montsalvens Arch Dam (Switzerland)
(LCH-EPFL)

Dynamic response of the drainage system of a cracked plunge pool liner due to free falling jet impact

M. Mahzari
Mahab Ghodss Consulting Engineers, Tehran, Iran

F. Arefi
Mahab Ghodss Consulting Engineers, Tehran, Iran

A.J. Schleiss
Laboratory of Hydraulic Constructions (LCH), Swiss Federal Institute of Technology, Lausanne (EPFL), Lausanne, Switzerland

ABSTRACT: The dynamic response of a drainage system of a concrete lined plunge pool due to loading by spillway water jet impact is investigated by use of a numerical model. Despite the primary assumption of a complete impervious concrete liner for the plunge pool, it is considered, that due to presence of cracks and joints in the liner, dynamic pressure of the water jet is propagated within the drainage system which, in turn, may result to further pressure amplification in the drainage pipes. A finite element formulation is used to simulate this phenomenon. Loading of system is generally defined as pressure time-history produced by the water jet, however, regarding to random nature of this loading, a stochastic analysis is employed taking a Power Spectral Density (PSD) function as the input loading. In this way, response of the system is evaluated in a probabilistic sense. The analysis revealed, that in some regions of the drainage network, amplification of significant order may occur. These high local pressure values can result to high dynamic uplift pressures which may cause local failure of the anchorage system and the concrete slab.

1 INTRODUCTION

Scouring of plunge pool bottom and walls due to impact of water jet coming from spillways, are of significant importance as it is a general problem observed in most of dams. This specially is a very critical matter when rock properties of plunge pool region are sensitive to scouring and deep scour holes have to be expected.

Scouring of a plunge pool near the dam foundation not only increases potential of sliding of valley slopes, but also may threaten abutments of the dam itself that in turn may end to hazardous situation for the dam stability. Therefore, severe measurements are to be taken to prevent excessive scoring of plunge pool and provide sufficient safety of stability for the valley slopes.

There are several methods to protect plunge pool against scouring. A method that is used in a number of dam projects is to fully line the plunge pool walls and bottom with a reinforced and anchored concrete liner.

However, the most sever loading that endangers operation of the liner system is the dynamic uplift pressure. If the liner is assumed to be impervious, uplift pressure normally depends on the tailwater elevation, downstream of the tailpond dam (if present). However, if the liner is pervious, dynamic pressure acting at the surface of the lining due to jet impact could be transferred underneath it resulting in very high uplift pressure that requires heavy and uneconomical anchorage system. Hence, it is of great importance that the liner is designed as an impervious liner. This, of course, requires high attention and special treatment of construction joints and sealing of the slab surface (for example coating with steel fibre high tensile concrete). By the use of an impervious liner, dynamic uplift pressure can be avoided and the anchor bolts limited to an acceptable amount. Furthermore to prevent high uplift pressure due to seepage during dewatering, usually a drainage system in form of a pipe and gallery network is located underneath the concrete liner at bottom of the plunge pool. By efficient use of pumps or free drainage, high uplift pressures are released so that they can be kept under a prescribed level.

However, there are some shortcomings associated with this system. First, there are construction joints in the liner system, which are normally sealed with double waterstops. Due to jet impact and induced vibration of the concrete slab, failure of the waterstops can not be excluded and the assumption of an impervious liner isn't valid anymore. It should be noted that in practice the repair of waterstops is not possible. If the waterstops fail in early stages of spillways operation, the joints remain unsealed for the rest of life of the system.

Moreover, it has to be noted that despite any special treatment for concrete surface, initiation and propagation of cracks through the concrete liner due to several reasons such as shrinkage or temperature variations can not be excluded completely. Accordingly, the assumption of full impervious liner for plunge pool during the design may not be valid in all cases. As a consequence the residual risk associated with this assumption should be investigated.

In the following, a specific study for the concrete lined plunge pool of KARUN III dam located in Iran, 460km south-west of Tehran is presented. KARUN III is actually under construction on KARUN River in KHOZESTAN province, mainly for development of hydroelectric power. The project comprises a 205 m high double arch dam, an underground powerhouse in the right abutment with an initial installed capacity of 2000 MW, flood releasing structures as orifice, free crest and ski jump spillway followed by a concrete lined plunge pool and a tailpond dam (Fig. 1).

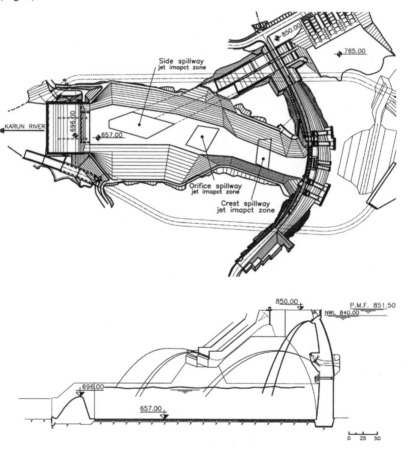

Figure 1. General layout and main vertical section of Karun III dam with plunge pool and tailpond dam.

228

The plunge pool of the dam was initially designed with a concrete liner as the side slopes had stability problems and depth of scour at the bottom was expected to be too high. Furthermore a drainage system was placed under the plunge pool lining with the purpose to release high uplift pressures due to seepage during dewatering.

2 PURPOSE OF THE ANALYSIS AND LAYOUT OF THE DRAINAGE SYSTEM

As already mentioned, there may occur some cracks or failure of the construction joint sealings in the plunge pool liner due to pressure fluctuations produced by impact of incoming water jets during operation of spillways. The dynamic pressure at the plunge pool bottom could be transferred through the cracks or broken joints into the drainage system. There these pressures may be amplified by the response of the drainage network system resulting in locally high pressures. The amplification is mainly due to reflection and superposition of pressure waves in the system that creates some standing wave patterns and resonance phenomena in the pipes. The response of the drainage system highly depends on the geometry of network and its boundary conditions. Therefore in the case of a cracked plunge pool liner, high dynamic pressures could occur in the drainage system that are much higher than the values considered for the design if the impervious lining. As a result, high local pressure created in the pipes can produce dynamic uplift acting on the liner and consequently result in failure of anchorage system or the liner itself. Therefore, it is very important to estimate amplification of water pressures in the drainage system due to a possible cracking of the liner and detect high pressure zones underneath the liner.

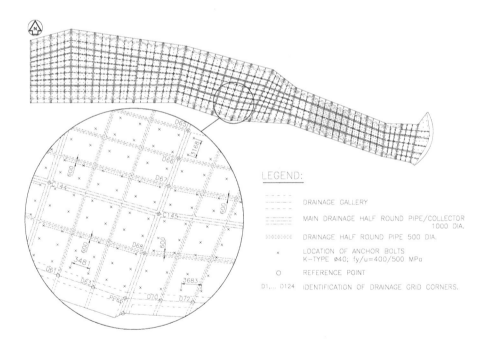

Figure 2. Layout of drainage system and anchors of the plunge pool of KARUN III dam.

The drainage system studied consists of two main galleries on both sides at the toe of the valley slopes, which are connected to tailwater. The water underneath the 3 m thick concrete lining is drained by a system of collectors (half pipes 500 mm diameter) and main drainage pipes (1000 mm diameter) guiding the water into the two side galleries (Fig. 2).

A comprehensive hydraulic model with scale of 1:80 was constructed at the Water Research Center in Tehran with the aim to study the general behaviour of the flood releasing systems. Pressure time-histories at several locations at the plunge pool bottom were measured for various releasing discharges from the spillways. These pressure time-histories were used for the design of the concrete lining of the plunge pool and will be discussed in the next chapter more in detail.

3 WATER PRESSURE FLUCTUATION RECORDS AND LOADING OF THE DRAINAGE SYSTEM

Pressure fluctuations transferred to the drainage system through a crack or a failed joint, depend on the pressure fluctuations created at the surface of bottom slab in the plunge pool due to jet impact. Generally, regarding to wave propagation nature of the phenomenon (water hammer), energy loss of waves travelling through the crack is not significant. Thus it may be assumed, that pressure fluctuations in the plunge pool are directly transmitted into the drainage system as an input dynamic loading. The main interest of the analysis is to have an idea about the possible amplification of the dynamic pressure in the drainage network. The effect of head losses in the crack in the concrete lining may be independently calculated and incorporated in the input loading. Therefore, it is important to determine pressure fluctuations at the plunge pool bottom accurately.

In the hydraulic model, pressure time-histories were recorded in several locations. Sampling rate of data in the model was $1/\Delta=1/0.005=200$ points/sec. According to Froude scaling, real time scale will be $\Delta=0.0447$ sec ($=\sqrt{80}\times0.005$). Regarding to this value of Δ, Nyquist frequency of this random process should be smaller than $f_c=11.180$ Hz to guaranty validity of the sampling (Press et al. 1997). However, recent studies (Bollaert & Schleiss 2001) indicate that pressure records of high velocity jet impact contain considerably higher frequency components that have a significant contribution in the motion

To check this, some investigation on the recorded time-histories in the model are performed. Figure 3 indicates some samples taken from the data ensemble.

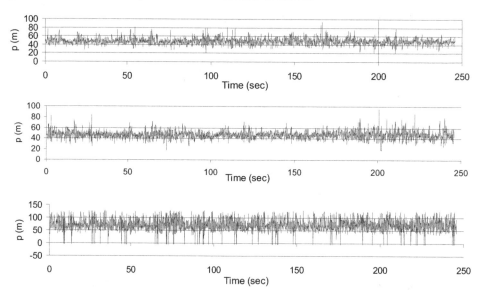

Figure 3. Pressure time-history samples in the plunge pool recorded in the hydraulic model.

To check the sampling rate, Fourier transformation (Press et al. 1992) of these records are calculated and plotted versus frequency in Figure 4.

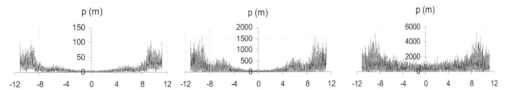

Figure 4. Fourier spectra of the samples of Figure 3.

It can be clearly seen, that sampling rate does not cover the frequency content of the motion and frequencies beyond f_c are aliased into the frequency range of $-f_c < f < +f_c$. Thus, use of these pressure records from physical hydraulic model for evaluation of input loading may not accurately estimate response of the system.

There are recent studies on this field and accurate measurements are performed to evaluate frequency content of these records for various conditions. Bollaert & Schleiss (2001 a, c, d) have conducted laboratory tests with prototype jet velocities to simulate pressure fluctuations at the plunge pool bottom. High sampling rate was used to catch all the frequencies that may be present. The results clearly indicate that high frequency components have remarkable contribution in the motion. Regarding the inaccuracy for the sampling rate of the hydraulic model, a Power Spectral Density (PSD) function, derived according to results represented by Bollaert (2001) and Bollaert & Schleiss (2001 a, d) were used. General form of the one-sided PSD function is shown in Figure 5.

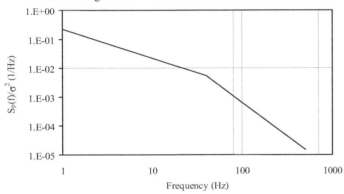

Figure 5. Power Spectral Density (PSD) function of pressure fluctuations at the plunge pool bottom.

It is possible to normalize the PSD function according the level of root mean square of the input motion. It was assumed, that all the records have zero mean, thus only dynamic fluctuation part of pressure records are considered in the analysis. Accordingly, dynamic pressure developed in the drainage system is obtained by the analysis, which can be superimposed to the hydrostatic pressure. Based on the assumed root mean square (standard deviation) response of the system is evaluated. This PSD function will be completely defined and used in post processing phase.

4 NUMERICAL MODEL AND ANALYSIS PROCEDURE

Behaviour of the fluid within the drainage system is governed by the general Navier-Stokes equations. Regarding to the small width of the cracks expected in the lining and reaching the drainage system and the highly fluctuating pressure, net flow into and out from the drainage system can not develop and one may assume that this flow is zero (averaged over time). Therefore, no significant flow in the drainage system is anticipated and hence, convection terms of

Navier-Stokes equations are removed and these equations are reduced to simple pressure wave equation (Kinsler et al. 1982):

$$p_{,ii} - \frac{1}{c^2}\ddot{p} = 0 \tag{1}$$

where summation over repeated indices is applied and p corresponds to the dynamic pressure part and c is the celerity or pressure wave velocity in the fluid.

With a finite element discretisation, Equation 1 is transformed to following discrete system of differential equations (Zienkiewicz & Taylor 1991):

$$\mathbf{m\ddot{p} + c\dot{p} + kp = F} \tag{2}$$

In this equation $\mathbf{m}$ is the mass matrix, $\mathbf{c}$ the damping matrix, and $\mathbf{k}$ the stiffness matrix of the finite element system and the vector $\mathbf{F}$ represents the external nodal forces. $\ddot{\mathbf{p}}$ and $\dot{\mathbf{p}}$ are the second and first derivative of nodal pressure values with respect to time.

Damping maybe introduced in the system due to several energy dissipating sources such as wall energy absorption, material damping or wave radiation from the system. If the external forces have a harmonic form of $\mathbf{F}=\mathbf{F_0}(\omega)e^{-i\omega t}$, response of the system is assumed to be of harmonic form with same frequency and a phase angle:

$$\mathbf{p} = \mathbf{p_0}(\omega)e^{-i\omega t} \tag{3}$$

where $\mathbf{p_0}$ is defined as

$$\mathbf{p_0} = (\mathbf{k} - i\omega\mathbf{c} - \omega^2\mathbf{m})^{-1} \times \mathbf{F_0} \tag{4}$$

As it can be seen, phase angle occurs only if source of damping is present in the system.

To solve the system for a given PSD, a frequency sweeping procedure is used. In this procedure, response of the system to a predefined pattern of loading with unit magnitude is evaluated over the frequency range of interest and Frequency Response Function (FRF) for each node is constructed. Since the system is linear, PSD of response of each node is derived by frequency response function of the node and PSD of input loading. This way PSD of response is determined and can be interpreted in a stochastic sense. However, this procedure has some limitations. The most important limitation is that the force distribution over the system is assumed to be fully correlated and relative correlation functions can not be defined in the analysis. The distribution of the force may be defined with various phase angles to somehow include this effect. However, explicit definition of correlation function is not implemented in the used code. The drainage system is modelled by 3D hexahedral solid elements to include local pressure amplifications. Geometry of the finite element model of the drainage system is shown in Figure 6.

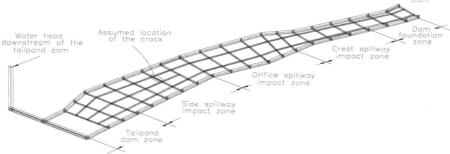

Figure 6. Finite element model of the drainage system and assumed location and length of the crack.

232

Low order (no mid-side node) elements have been used with standard shape functions. Mesh size is so selected to provide enough number of divisions along a wavelength (6~10 number of divisions along a wavelength).

The end vent of the drainage system, which extends into the tailpond dam is modelled and zero pressure boundary condition is applied to its free surface. No effect of surging of the free surface was considered in the analysis, which is supposed to have no significance effect on the response of the pipe network.

The location of the crack was assumed in the jet impact zone of the ski jump spillway. The crack across the concrete lining reaches directly the drainage system. The crack is assumed to be linear and the loading is applied over the nodes located beneath the assumed crack location. The loading is applied as unit pressure on these nodes (i.e. in form of primary DOF excitation).

In principal different locations of crack and its amplification of pressure waves in the system should be investigated. Furthermore the interaction of different cracks at several impact zones in the plunge pool could be studied in a more detailed analysis.

5 DISCUSSION OF RESULTS

The post-processing of the results is the most time consuming stage of the whole analysis process. As already mentioned, the frequency response function of each individual node is derived in the analysis phase. The frequency response function and input loading PSD function are used to evaluate PSD function of pressure response in the node under consideration. By use of response PSD, pressure root mean square for each node is evaluated (Newland 1996). Probabilistic characteristics of response of each individual node are then fully defined assuming a normal probability distribution for the output motion, same as what is assumed for the input.

To evaluate frequency response function, a frequency increment of 1.0 Hz was used to increase precision of the results.

For post-processing purpose, the parameters of input PSD function (Fig. 5) have to be determined. By assuming a root mean square of 100 kPa for the input loading and maximum frequency of 100 Hz, the so defined parameters of the input PSD function are represented in Table 1 and shown in Figure 7.

Table 1. Parameters of input PSD function used for the analysis

Frequency (Hz)	1.00	40.00	100.00
Spectral Density Value (kPa^2/Hz)	2370.883	59.272	6.988

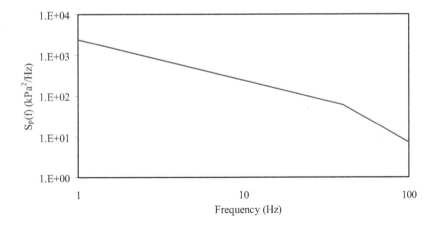

Figure 7. Graphical PSD function with a RMS of 100 kPa used for analysis.

Note that this function is a one-sided PSD function and area below this function represents mean square of the input motion. Compared to the PSD function given by Bollaert (2001) and Bollaert & Schleiss (2001 a, d), the curve is truncated for frequencies higher that 100 Hz. This is mainly because of insignificant contribution in the motion of frequencies higher than 100 Hz and numerical difficulties for high frequency analysis.

In a node by node basis, PSD of input (Fig. 7) and frequency response function are combined to derive output pressure PSD. From the output PSD root mean square at each node is evaluated. Plotting pressure values for each node, location and value of pressure amplification is detected through the drainage system.

Response of the system depends on several parameter including pipe network geometry, input loading and properties of the fluid. Pressure wave velocity or celerity, c, is one of the main parameters which may affect level of amplification in the system. For water, pressure wave velocity is about 1430 m/s in a complete rigid system. However for the drainage system this value may not be valid.

Beside the flexibility of the drainage system, the existence of air bubbles in the system coming from the plunge pool during operation of the spillways can reduce pressure wave velocity (Bollaert & Schleiss 2001 a).

Accordingly, the response of the system was analysed for different values of celerity, namely 700 to 1400 m/s with and increment of 100 m/s. The one-sigma results (root mean square) of dynamic pressure are shown in Table 2 and figure 8. In a probabilistic sense, the values of these results have a probability of exceedence of 31.7%. Moreover, assuming a log-normal probability distribution for the motion amplitude, probability of exceedence of amplitude from the values shown in Table 2 is 60.7%.

Table 2. Maximum dynamic pressure (probability of exceedence = 31.7 %) in different zones of the plunge pool (Fig. 6) for various celerity values.

Wave celerity of water (m/s)	Maximum dynamic pressure in plunge pool zone (in kPa)				
	Tailpond dam	Side spillway	Orifice spillway	Crest spillway	Arch dam foundation
1400	370	560	500	560	560
1300	784	697	348	435	522
1200	355	355	240	240	240
1100	195	290	290	195	195
1000	238	358	358	318	278
900	373	373	207	207	165
800	260	390	170	170	260
700	466	700	310	310	310

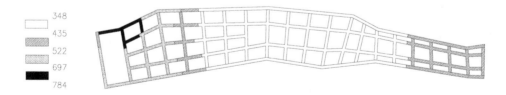

Figure 8. One-sigma dynamic pressure distribution for celerity value c=1300 m/s (probability of exceedence = 31.7 %).

In the conducted analyses, a slight damping is associated with energy absorption of walls of the pipe network that the value of absorption coefficient is taken to be μ=0.01% (Craggs 1980). This will prevent the response singularity when the exciting frequency approaches eigenvalues of the system.

Table 2 depicts maximum dynamic pressure (in kPa) in the drainage system for various celerity values. As it is seen, high pressure zones are created in some zones, where the risk of dynamic uplift pressure and failure of the concrete slab is increased compared to the input motion that is taken as a random function with root mean square (RMS) of 100 kPa, local amplification are observed for all cases. An amplification factor is defined that is ratio of RMS of the pressure history in each node to the RMS of the input loading (100 kPa).

For the case of water with no air content (c=1400 m/s) a maximum amplification of about 5.6 is calculated. From Table 2, it can be seen, that high pressure values occur within a large portion of the main galleries near the dam. This is of critical importance for the plunge pool as these high pressure values may further affect stability of the liner that is constructed for stability of the side walls.

For the celerity of 1300 m/s almost identical pressure values occur within the main galleries, however, in this case high amplification near the tailpond dam with an order of about 8 has taken place (Table 2 and Fig. 8). Considering the former case some slight increase in pressure values could be seen near the tailpond dam where the values were not so high and it is clearly seen that for the second case pressure at these zones are highly increased. However, it should be noted that the zones with high amplification are localised near the tailpond dam.

For the third case (c=1200 m/s), maximum amplification is detected within the collector pipes and the value of amplification is almost halved. A uniform pressure distribution is present in the main galleries with some slight raise at vicinity of the dam body and the tailpond dam. For this case a maximum magnification of 3.5 has taken place and rapid variations of pressure in the collectors are generated. Again it is seen that location of maximum pressure is near the tailpond dam.

Analysis results with the celerity of 1100 m/s indicate the same behaviour as the former case and the maximum pressure is further decreased. As well, location of the maximum pressure is near the assumed crack in the lining.

For the celerity of 1000 m/s, almost the same magnification takes place in the main galleries and the collectors. Pressure variations along the main galleries are more intense and approximate location of the maximum pressure is near the assumed location of the crack in the plunge pool lining. For this case a magnification of about 3.5 is calculated.

For the next two cases (c=900 and c=800 m/s), an almost identical pressure distribution is obtained. Pressure variations in the main galleries are not high and slight raise at both ends of them are observed. Maximum value of pressure occurs in the collectors with a magnification factor of 3.7~3.9. By decrease of celerity, slight increase in amplification has occurred.

By further decrease of celerity to c=700 m/s a jump in magnification factor of pressure is observed which is about 7.0. Where for the latter two cases amplification took place near the tailpond dam, for this celerity value, high pressure near the assumed location of crack is experienced. However, pressure in main galleries is in same order as the former cases.

Regarding to the presented discussion on results, one should note that the pressure distribution in the drainage system highly depends on geometry of the pipes, damping and wave celerity of water. In view of the high uncertainty of the wave celerity value of aerated water due to working of the spillway and penetration of air bubbles through the cracks into the drainage system pipes, detection of exact location of pressure amplification is quite difficult. Moreover, regarding to the effect of damping on the response of the system, the damping value associated with the wall energy absorption is difficult to determine.

The same calculations were repeated for a higher value of damping (μ=0.1%). The results (not shown here) indicated, that pressure amplification is still observed in the drainage system. Moreover, in all cases, maximum location of pressure was observed near the crack location in collector pipes and no significant amplification in main galleries was achieved. However, amplification order was reduced and for the celerity range used in the analyses, amplification factor was between 1.3 and 2.0.

It should be noted that the analysis does not include effect of surging of water at vents as a fixed boundary was assumed for the system. Surge vents can decrease pressure wave amplification in the drainage systems to some extent. However, it is expected that they may have only limited effects.

In order to analyse also the spatial distribution of pressure, cumulative distribution of pressure in the drainage system was calculated for all the celerity values given in Table 2. To illustrate this cumulative distribution, the part of the whole length of the drainage system, where the pressure is less than the specified value Ps, was determined for various celerity values in Figure 9.

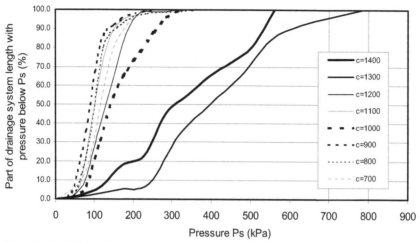

Figure 9. Spatial distribution of pressure in the drainage system as a function of wave celerity c (m/s).

The vertical axis of Figure 9 shows the part (0 to 100%) of the total length of the drainage system for which the dynamic pressure is smaller than a specified value in the horizontal axis. It can be seen, that for low bubble or air content, that means high wave celerity in the drainage system, significant deviation is observed in the spatial pressure distribution. If the compressibility of water is reduced due to increase of air content in water, the deviation in spatial distribution is reduced and pressure is more localised in the system.

6 CONCLUSIONS

The response of a typical drainage system under a concrete lined plunge pool was investigated. The response was analyzed assuming a single crack in the lining where pressure waves due to water jet impact can penetrate into the drainage system. The analyses focussed on the pressure amplification as a result of the creation of standing and resonance pressure wave pattern in the drainage system. The analysis revealed amplification factors greater than 5 to 8 in some local regions in the drainage system.

The sensitivity of the pressure wave celerity analysis as a function of the air bubbles content in water in the drainage system and its deformability was studied. The analysis indicates that a decrease of the pressure wave celerity, slightly increases the dynamic pressures in the drainage system and the zones of the amplification of the pressure are changed. It may be concluded, that even carefully drained concrete linings of plunge pools may be loaded by high dynamic uplift pressures, if the cracking of the lining can not be excluded.

REFERENCES

Bollaert, E. 2001. Spectral density modulation of plunge pool bottom pressure inside rock fissures, Proceedings of the XXIXth IAHR Congress, Beijing.

Bollaert, E. & Schleiss, A. 2001a. Air bubble effects on transient water pressures in rock fissures due to high velocity jet impact, Proceedings of the XXIXth IAHR Congress, Beijing.

Bollaert, E. & Schleiss, A. 2001b. A new approach for better assessment of rock scouring due to high velocity jets at dam spillways, Proceedings of the 5th ICOLD European Symposium, Geiranger, Norway.

Bollaert, E. & Schleiss, A. 2001c. Scour of rock due to high velocity plunging jets. Part I: a state-of-the-art review, submitted to the Journal of Hydraulic Research, IAHR, Delft, The Netherlands.

Clough, R.W., J. Penzin (2nd ed.) 1993. Dynamics of Structures. Singapore: McGraw-Hill, Inc.

Craggs, A. 1980. A Finite Element Model for Acoustically Lined Small Rooms. *Journal of Sound and Vibration* 108(2): 327-337.

Kinsler, E.L. et al. 1982. Fundamentals of Acoustics. New York: John Wiley and Sons.

NAFEMS (1989). Selected Benchmarks for Forced Vibration. Report prepared by W.S. Atking Engineering Sciences.

Newland, M (3rd ed.)1996. Random Vibrations, Spectral and Wavelet Analysis. Singapore: Longman.

Press, W.H., S. A. Teukolsky, W.T. Vetterling, B.P. Flannery (2nd ed.) 1997. Numerical Recipes in C, The Art of Scientific Computing. Cambridge: Cambridge University Press.

Thomson, W.T. (2nd ed.) 1983. Theory of Vibration with Applications. London: George Allen & Unwin.

Vierck, R.K. (2nd ed.) 1979. Vibration Analysis. New York: Harper & Row Publishers.

Zienkiewicz, O.C., R. Taylor 1987. The Finite Element Method, Vol. I. London: McGraw-Hill, Inc.

Zienkiewicz, O.C., R. Taylor 1991. The Finite Element Method, Vol. II. London: McGraw-Hill, Inc.

Genetic algorithm optimization of transient two-phase water pressures inside closed-end rock joints

E. Bollaert
Laboratory of Hydraulic Constructions, Swiss Federal Institute of Technology, Lausanne, Switzerland

S. Erpicum
Laboratory of Applied Hydrodynamics and Hydraulic Constructions, University of Liège, Belgium

M. Pirotton
Laboratory of Applied Hydrodynamics and Hydraulic Constructions, University of Liège, Belgium

A.J. Schleiss
Laboratory of Hydraulic Constructions, Swiss Federal Institute of Technology, Lausanne, Switzerland

ABSTRACT: High-velocity plunging water jets, appearing at the downstream end of dam weirs and spillways, can create scour of the rock. The prediction of this scour is necessary to ensure the safety of the toe of the dam as well as the stability of its abutments. A physically based engineering model has been developed at the Laboratory of Hydraulic Constructions for evaluation of the ultimate scour depth. This model is based on experimental measurements of water pressures at plunge pool bottoms and inside underlying rock joints. The pressures inside the joints revealed to be of highly transient nature and governed by the presence of free air. Hence, a numerical modelling of these pressures was performed, in collaboration with the Laboratory of Applied Hydrodynamics and Hydraulic Constructions (HACH), based on the one-dimensional transient flow equations applied to a pseudo-fluid. The amount of free air is a function of the instantaneous pressure inside the joint and has been accounted for by means of appropriate celerity-pressure relationships. These relationships are defined by the ideal gas law and Henry's law and were optimised by means of a genetic algorithm optimisation technique. Very good agreement has been obtained between the measured and computed pressures at the end location of one-dimensional closed-end rock joints.

1 INTRODUCTION

High-velocity plunging water jets, appearing at the downstream end of dam weirs and spillways, can create scour of the rock. Scour is often predicted by empirical or semi-empirical formulae, developed from physical models or prototype observations. These formulae are not fully representative because they cannot describe all of the physical effects involved. Above all, the characteristics of pressure wave propagation in the fissures of the jointed rock mass are unknown. Therefore, an experimental facility has been built at the Laboratory of Hydraulic Constructions that measures pressure fluctuations at plunge pool bottoms and simultaneously inside underlying rock joints (Figure 1).

The facility simulates the falling jet, the plunge pool and the fissured rock mass. Prototype jet velocities are impacting through a water cushion onto a plunge pool bottom, generating so pressure fluctuations that can enter the underlying rock joint. The water pressures are measured simultaneously at the pool bottom and inside the simulated rock joint. As such, a direct relationship between pressures in plunge pools and pressures that are transferred inside a jointed rock mass has been outlined. A more detailed description of the facility can be found in Bollaert & Schleiss (2001a) and Bollaert (2002).

The water pressures that were measured inside a one-dimensional I-shaped rock joint (Figure 2) have been computed numerically. For this, the one-dimensional transient flow equations have been applied to a pseudofluid with a homogeneously distributed free air content. The exact free air content thereby depends on the instantaneous pressure value inside the joint and is expressed by means of "wave celerity-pressure" relationships. These relationships follow two basic physical laws: the ideal gas law and Henry's law. In the following, emphasis is given on the optimisation of the parameters of these relationships. As a first step, a trial-and-error optimisation of the parameters has been performed by comparing the experimentally and numerically obtained maximum and minimum pressure values as well as the root-mean-square value of the pressure fluctuations. Secondly, a more objective optimisation has been performed based on the use of genetic algorithms. A very good agreement was obtained with the experimentally measured pressure values.

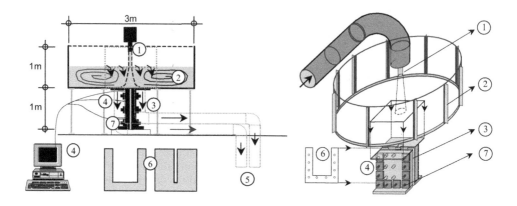

Figure 1. Side view and perspective view of the experimental facility: 1) cylindrical jet outlet, 2) reinforced plastic cylindrical basin, 3) pre-stressed two-plate steel structure, 4) PC-DAQ and pressure sensors, 5) restitution system, 6) thin steel sheeting pre-stressed between steel structure (defining the form of artificial one-and two-dimensional rock joints), 7) pre-stressed steel bars.

2 EXPERIMENTAL MODEL

The experimentally measured water pressures have been recorded by a series of pressure sensors. Two of these sensors are of particular importance: sensor (a) is located at the plunge pool bottom, close to the rock joint entrance and quasi under the jet's centerline, while sensor (d) is located at the end position of an I-shaped closed-end rock joint (Figure 2). Pressure measurements have been made simultaneously at these two locations, in order to point out a direct relation between the two locations.

This comparison revealed that the water pressures inside the rock joint are governed by the propagation, superposition and reflection of pressure waves induced at the entrance. These waves revealed a transient and cyclic behavior, which is defined by the presence of free air. Two physical laws describe this: the ideal gas law and Henry's law. The former expresses the change in volume of a free air bubble as a function of pressure. The latter describes the quantity of air that is released from the liquid or dissolved in the liquid due to a change in pressure. These two laws result in "wave celerity-pressure" relationships that express the non-linear transient characteristics of the system.

The water pressures have been measured inside four closed-end joints and one open-end joint. The transient pressures in an I-shaped rock joint are dealt with here in particular. These are characterized by a continuous change between peak pressures and periods of low, near-atmospheric pressure. The frequency of these changes is governed by the amount of free air. Peak pressures up to several times the kinetic energy head of the impacting jet (= $V^2/2g$) have been measured, indicating the formation of standing waves and resonance conditions. Free air contents were between 0.5 and 10 %, corresponding to wave celerities ranging from only 50 to 250 m/s.

240

The measurements at sensor (a) have been used as weak upstream boundary condition for the numerical calculations of the corresponding pressures inside the closed-end rock joint. The so obtained pressures are so compared with the experimentally measured ones.

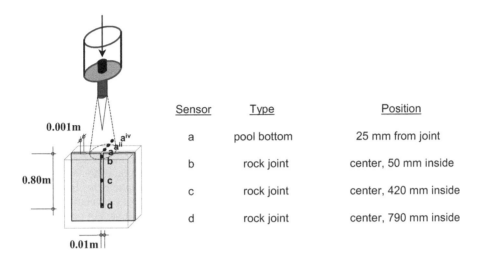

Sensor	Type	Position
a	pool bottom	25 mm from joint
b	rock joint	center, 50 mm inside
c	rock joint	center, 420 mm inside
d	rock joint	center, 790 mm inside

Figure 2. Schematization of pressure sensor locations at the plunge pool bottom and inside the I-shaped rock joint. The pressures measured at sensor (a) are used as upstream boundary condition for the numerical calculations. At sensor (d), a comparison is made between the experimentally measured pressures and the numerically calculated ones.

3 NUMERICAL MODEL

3.1 The model equations

The numerical simulation makes use of the one-dimensional transient flow equations for a homogeneous two-component air-water mixture. These are as follows (Bollaert et al., 2001c):

$$\frac{\partial p}{\partial t} + \frac{c^2}{g} \cdot \frac{\partial V}{\partial x} = 0 \qquad \text{mass conservation} \qquad (1)$$

$$\frac{\partial(\eta V)}{\partial t} + \frac{\partial}{\partial x}\left(\beta V^2\right) + g \cdot \frac{\partial p}{\partial x} + \frac{1}{2} \cdot \frac{\lambda}{D} \cdot V \cdot |V|^e = 0 \qquad \text{momentum conservation} \qquad (2)$$

in which p is the pressure head (m), V the mean velocity (m/s), c the pressure wave celerity (m/s) and D the hydraulic diameter. For a small joint, the hydraulic diameter D may be taken equal to twice the joint thickness. The terms λ, η and β account for steady, unsteady and non-uniform velocity distribution friction losses. They are three parameters to be optimized. The steady friction factor λ is calculated based on the Colebrook-White formula. The unsteady friction factor η depends on the cyclic behavior of the flow inside the joints. For a uniform velocity distribution, β equals 1 in the convective term of equation (2).

It is assumed that these friction terms also incorporate other possible energy losses, such as friction due to heat or momentum exchange between the air and the water phase. As such, they cannot be compared with the Darcy-Weisbach friction term that is usually applied for one-phase steady-state flow. Their values are often quite different, due to the particular damping effect generated by the two-phase transient character of the flow (Martin & Padmanabhan, 1979;

Ewing, 1980). For turbulent flow conditions inside the joint, the exponent e has to be taken equal to 1. However, as a result of the narrow geometry, the Reynolds numbers can be very low ($\sim O(10^2)$) and laminar flow might be more plausible under certain circumstances. The corresponding exponent e has then to be taken equal to 0.

The two-component air-water mixture inside the joint is simulated as a pseudo-fluid with average properties and, thus, only one set of conservation equations. The density is hardly modified by the gas and, at relatively small gas contents, may be approximated by the density of the liquid. This means that any possible mass or momentum transfer between the two components is excluded. Furthermore, no slip velocity or heat transfer between the two phases is considered, so the energy equation is omitted. According to Wylie & Streeter (1978), this simplified approach is valid for air contents of at least 2 %. Martin & Padmanabhan (1979) numerically verified the homogeneous flow assumption for air contents of up to 30 %, and found correct wave celerities. Therefore, in the here presented approach, the homogeneous flow model has been applied. No further assumption is made regarding the distribution of air throughout the joint, so the wave celerity c is dependent on both time and space.

During the experiments, the air concentration inside the joints was found to depend on the pressure. The volume of the air bubbles continuously changed as a function of the latter. Thus, significant transfer between the two phases occurs. Two physical laws dictate this transfer: the ideal gas law and Henry's law. This assumption was based on the shape of the celerity-pressure relationships, which were found to be of changing mass of free air as a function of pressure.

Hence, it has been preferred in the following to verify this basic assumption by using a constitutive relationship between the celerity $c(x,t)$ and the pressure $p(x,t)$. This relationship replaces any kind of transfer (heat, mass or momentum) that could occur between the air and the water and has the advantage of simplicity. It is dependent on both space and time. A quadratic form seems to match quite well with the measured data points and is written as follows:

$$c(x,t) = k_1 + k_2 \cdot p(x,t) + k_3 \cdot p^2(x,t) \tag{3}$$

in which k_1, k_2 and k_3 are three numerical parameters that have to be optimized. In some cases, a double quadratic form revealed to be more appropriate.

3.2 The discretization and numerical scheme

The numerical scheme that is used to solve a weak formulation of this set of three equations is a 2nd order finite-volume scheme. As the experimental pressure measurements revealed the appearance of violent transient and highly non-linear wave phenomena, it is obvious that a shock-capturing scheme, introducing a fit amount of numerical dissipation without excessive smearing of the peak pressures, is preferable.

The numerical code defines an unsteady pressure signal as weak upstream condition and imposes a zero flow velocity as weak downstream condition (at the end of the joint). The upstream pressure signal has been taken from the experimental measurements made at the entrance of the rock joint. A weak formulation of the upstream boundary condition has been chosen. This implies that the upstream boundary condition $p_{up}(t)$ is only applied as outer condition on the upstream finite volume, and not as condition over the whole volume directly. It is believed that this smoothing of the boundary condition is most plausible for the physical situation. The boundary conditions have been presented in Figure 3 together with the numerical grid. The total length of the I-joint of 0.80 m has been discretized into 40 elements of 0.02 m each. This gave an optimum equilibrium between precision and time duration of the calculations.

Furthermore, an adaptive time stepping has been applied. The criterion that has been used to determine the critical time step is a classical Courant condition, in which the Courant number $C_t = (V+c) \cdot (\Delta t / \Delta x)$ is taken equal to 0.5. This condition is checked at every node of the system, and the most restrictive one is retained for the next time step. The calculations revealed numerical time steps on the order of $1 \cdot 10^{-5}$ to $5 \cdot 10^{-5}$ seconds, i.e. one to two orders of magnitude smaller than the time step of the experimental tests ($= 0.001$ sec at 1'000 Hz acquisition rate).

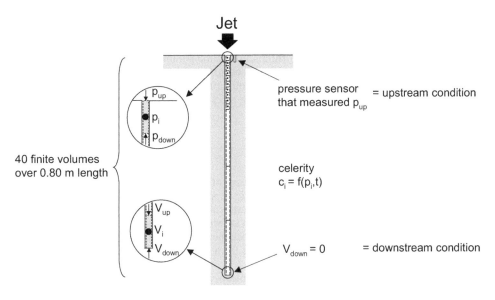

Figure 3. Definition of the numerical grid and of the upstream and downstream boundary conditions.

3.3 Optimization criteria

The adjustment of the friction losses parameters λ, η, β, and of the three k-parameters, can be performed based on the following criteria: mean pressure value, root-mean-square value, maximum and minimum pressure values and finally the histogram and the power spectral density of the computed pressure values. The computed values are systematically compared with the corresponding measured values for different test runs.

For the histogram and the power spectral density, a least-square criterion has been applied. The application of such a criterion to the pressures in the time domain was found to be impossible due to the small time lag that can exist between the input pressure signal (measured at sensor (a)) and the output signal measured at the end of the joint (at sensor (d)). This time difference can significantly increase the sum of the least squares, even for an appropriate numerical solution. A solution to this problem has been searched by proposing a time decay parameter between the measured input pressure signal and the calculated pressures inside the joint. The calculations have shown, however, that this time decay is not a constant but continuously changes during the run. Hence, it proved unsatisfactory to solve the problem. However, when considering the histogram and power spectral density of the pressures, this time lag problem vanishes.

The optimization process has been performed for pool depths ranging from 0.20 m (core jet impact) to 0.67 m (developed jet impact), and for jet outlet velocities V_j between 10 and 30 m/s. Test run periods were of 10 seconds per optimization. This time period was found to procure an appropriate balance between correct numerical analysis and an acceptable computation time. However, for every run, the first second of calculations has been systematically omitted from the optimization process, in order to avoid influences of the initial condition. Preliminary tests with a sinusoidal input signal have shown that most of these influences die out after some tenths of seconds.

The optimization is characterized by two stages. In a first stage, a trial and error process has been applied. Based on the measured data, an appropriate range of values could be found for the three k-parameters. Within this range, an optimum was then searched for by performing several consecutive numerical runs and by comparing the mean, root-mean-square, maximum and minimum pressure values, as well as the obtained histogram. Although this approach is of subjective character, it allows defining the major tendencies of the celerity-pressure relation-ships.

The second stage of the optimization process of celerity-pressure relationships involves an automated process based on genetic algorithms. All of the parameters could so be optimized in an objective manner. Two criteria have been used to optimize genetically: the histogram of pressure values and the power spectral density. Both criteria gave very similar results.

These two phases of optimization are dealt with more details hereafter.

4 TRIAL AND ERROR OPTIMIZATION

4.1 Basic principle

For closed-end rock joints, a first determination of the celerity-pressure relationships has been performed by assuming a linear or a quadratic dependence between the two parameters. At relatively low celerities and pressures, one quadratic curve is sufficient to describe the relation-ship (optimisation of parameters k_1 to k_3). At high celerities and pressures, sometimes this same curve could still be used, but sometimes it was necessary to determine a second quadratic curve (optimisation of parameters k_4 to k_6). The change from one curve to the other is determined by means of a pivot celerity c_{PIV}.

The optimisation criteria are the mean, root-mean-square and extreme pressure values, as well as the histogram of pressure values. The optimisation has been done by performing succes-sive runs of the numerical model and by trial and error comparison of the so obtained results. As such, the followed optimisation procedure is of subjective character but nevertheless pro-cures a good idea of the global form of the appropriate celerity-pressure relationships. Table 2 summarizes the results for the considered range of experimental parameters. The jet velocities that are marked with * indicate results for a convergent jet outlet. The other jet outlets are of cylindrical shape.

Table 2: Parametric results of the preliminary optimisation applied to a 10 seconds test run for different jet velocities and plunge pool depths. Test runs that are marked with * are for a convergent jet outlet.

V_j m/s	Y m	k_1 -	k_2 -	k_3 -	c_{PIV} m/s	k_4 -	k_5 -	k_6 -	Flow -	λ -
14.7	0.60	45	12.5	0	-	-	-	-	turbulent	0.30
14.7	0.20	50	13.0	0	850	91.7	14.8	-0.04	turbulent	0.30
14.7	0.20	55	12.0	-0.024	-	-	-	-	laminar	0.20
14.7*	0.20	70	14.0	0	-	-	-	-	turbulent	0.80
14.7*	0.20	70	14.0	0	-	-	-	-	laminar	0.50
19.7	0.67	10	14.0	0	960	424	9.6	-0.024	turbulent	0.30
19.7	0.67	80	12.0	-0.024	-	-	-	-	laminar	0.50
24.6	0.67	15	12.5	0	800	37.2	14.4	-0.038	turbulent	0.50
24.6	0.67	55	11.8	-0.025	-	-	-	-	laminar	0.60
24.6	0.60	10	12.0	0	278	-15	13.8	-0.035	turbulent	0.50
24.6	0.60	20	11.0	-0.025	-	-	-	-	laminar	0.35
29.5	0.60	35	3.5	0	-	-	-	-	turbulent	0.50

The corresponding celerity-pressure relationships are visualized in Figure 4. Figure 4a pre-sents the combined linear or quadratic curves valid under turbulent flow assumptions. Figure 4b presents similar results but for laminar flow assumptions. Although the curves for turbulent flow seem to indicate slightly higher wave celerities, both cases exhibit similar tendencies.

The first optimisation started with the assumption of turbulent flow conditions. The results were accurate in terms of mean, root-mean-square and extreme pressures, but the calculated histogram of pressure values could not be fully adjusted to the measured histograms. When adjusted for minimum and maximum pressures, the intermediate pressure values didn't correspond at all. Similarly, when adjusting these intermediate values, the extreme pressures didn't match anymore. This phenomenon is probably due to a different friction at different pressure values. Moreover, the calculated flow velocities were generally very low, less than 0.2 m/s. For the narrow joint thickness of only 1 mm, this theoretically results in Reynolds numbers less than 100 and, thus, very laminar flow.

Therefore, it was decided to repeat the optimisation but for laminar flow assumptions. This resulted in a much better agreement of the calculated and measured histograms, as can be seen by comparing Figure 4b with 4d.

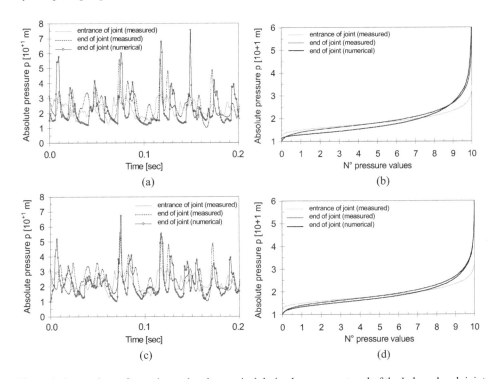

Figure 4. Comparison of experimental and numerical derived pressures at end of the I-shaped rock joint:
a) pressure signal for turbulent flow;   b) histogram for turbulent flow;
c) pressure signal for laminar flow;   d) histogram for laminar flow.

The measured data points were obtained for test runs with a high acquisition rate (5'000 to 10'000 Hz) and show good agreement with the numerically obtained relationships for jet velocities less than or equal to 25 m/s (Figure 5). The scatter of data at high jet velocities (> 20 m/s) is probably due to a varying air concentration in the plunge pool. This variation induces changes in free air content inside underlying rock joints. As a result, during a complete test run, different celerity-pressure relationships have to be accounted for.

This, however, cannot be taken into account by the numerical simulation. The phenomenon can be easily observed for a jet velocity of 29.5 m/s. Some of the data points match very well with the optimised celerity-pressure curve; other points exhibit a completely different air concentration. The celerity-pressure relationships have been optimised based on test runs at only 1'000 Hz of acquisition rate. This reduces the accuracy of the results, especially at higher pressures and celerities. This can be observed by an increasing discrepancy between the measured data points and the calculated curves at higher pressures. Hence, the celerity-pressure relationships at the higher pressure range should be interpreted with a lot of precaution. It is not excluded that they are not accurate.

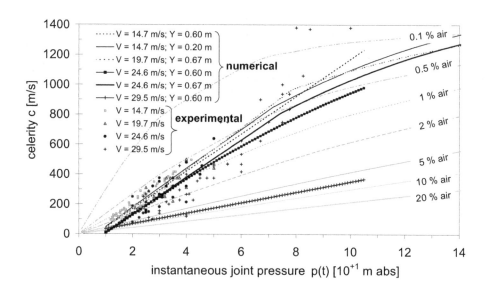

Figure 5. Numerically derived celerity-pressure relationships and comparison with measured data points for different jet velocities and plunge pool depths for turbulent flow assumptions.

For a jet velocity of 29.5 m/s and for turbulent flow assumptions, a totally different celerity-pressure relationship is obtained in Figure 5. It is the only curve that exhibits a constant mass of free air with increasing pressure. This mass corresponds to a standard air concentration of between 5 and 10 %. The other curves indicate a significant decrease of the mass of free air with increasing pressure, which can be described by Henry's law.

5 GENETIC ALGORITHM OPTIMIZATION

5.1 Basic principle

Genetic algorithms are exploration algorithms based on the natural selection and genetic mechanisms. They work on a set of N coded representations of the value of the parameters to optimize called a population of N chromosomes or chains. The performance of each chromosomes is evaluate thanks to a fitness function. The algorithm creates new populations of solutions from the previous ones using genetic operators such as selection, mutation and crossover, in order to find the optimum. The selection criteria are such that only the best individuals are used to built a new population, thus, a sort of natural selection is occurring.

It is obvious that genetic algorithms are situated between classical optimisation methods, applied locally in the search space of solutions, and purely random or systematic optimisation methods, that explore all possible solutions. Genetic algorithms constitute a sort of intelligent and pseudo-random exploration of the search space of solutions. Thus, they can be used to study discontinuous and disjointed functions and always converge on the absolute optimum.

On this basis, an optimisation module, available for a global application of the Wolf codes, was set up in the HACH (Belgium) in the framework of a Master Thesis (Erpicum, 2001).

This optimisation module has been used for this study using as the fitness function a simple least-square function between the measured and computed values of the histogram or the power spectral density.

5.2 Jet and plunge pool parameters

Comparison is made with the pressure measurements realized for the one-dimensional I-joint. This comparison is done at the joint end, i.e. at sensor (d).

The genetic algorithm optimization has been performed for the plunge pool depths and jet velocities as indicated at Table 3. The jet outlet was cylindrical, for a diameter of 72 mm.

Table 3: Plunge pool depths and jet velocities of the test runs that were used to compare with the numerical model.

| Jet velocity V_j | Plunge pool depth Y | | |
	0.20 m	0.60 m	0.67 m
9.8 m/s	x		x
14.7 m/s	x	x	x
19.7 m/s	x		x
24.6 m/s	x	x	x
27.0 m/s			x
29.5 m/s	x	x	x

Small plunge pool depths generate jets for which the core still exists and directly impacts on the rock joint. For pool depths higher than 0.50 m, a fully developed turbulent shear layer develops at the water-rock interface. Most of the experimental runs have been acquired at a rate of 1'000 Hz. When another acquisition rate was used, this has been explicitly mentioned.

5.3 Numerical optimization for a jet velocity of 19.7 m/s and a pool depth of 0.67 m

The procedure that has been followed is outlined in detail for a jet velocity of 19.7 m/s and a plunge pool depth of 0.67 m. As a first approach, the following parameters have been optimised: k_1, k_2, k_3, c_{PIV}, k_5, k_6, λ and η. This means that the correction term β that accounts for an uneven velocity distribution has been omitted. The parameter k_4 is not strictly necessary when knowing the pivot celerity c_{PIV}. Furthermore, the second quadratic curve as well as the pivot celerity c_{PIV} is difficult to optimise when no high-pressure values are available. An attempt was made but revealed to procure unsatisfactory results.

The used optimisation criteria are the histogram of the measured pressure values and the power spectral density of the fluctuating part of the measured pressures. The results of the genetic algorithm applied to a single test run of 10 seconds are summarized at Table 4.

Table 4: Parametric results of the genetic algorithm optimisation applied to a 10 seconds test run with a jet velocity of 19.7 m/s and a plunge pool depth of 0.67 m. The best results for are in bold.

| Function | Time | k_1 | k_2 | k_3 | c_{PIV} | k_4 | k_5 | λ | η | criterion |
-	sec	-	-	-	m/s	-	-	-	-	-
12'972	10	38	6.8	0.100	-	-	-	0.50	-	histogram
13'080	10	14.7	8.9	0.053	-	-	-	0.50	-	histogram
13'306	10	28.7	8.9	0.053	-	-	-	0.49	-	histogram
13'417	10	28.7	9.5	0.040	-	-	-	0.49	1.09	histogram
13'496	**10**	**22**	**10.2**	**0.047**	**-**	**-**	**-**	**0.53**	**1.14**	**histogram**
12'197	10	75.3	6.3	0.120	-	-	-	0.58	1.0	spectrum
12'236	**10**	**70**	**5.0**	**0.143**	**-**	**-**	**-**	**0.58**	**1.0**	**spectrum**

The "function" value is the parameter that has to be optimised by the least-squares criterion. The better the result, the higher is this value. The optimisation of the second quadratic curve, i.e. the parameters c_{PIV}, k_4 and k_5, proved unsatisfactory and is not presented. The friction coefficient λ is very close to the preliminary optimised value of 0.50, for both histogram and spectrum optimisations. The shape of the two best-fit quadratic curves (values indicated in bold at

Table 4) is presented in Figure 6. The two curves are of similar shape and in good agreement with the previously established optimum. The corresponding pressure signal in the time domain as well as the histogram and power spectral density are presented in Figure 7.

The numerically obtained power spectral content, however, overestimates the higher frequencies, due to the absence of thermal dissipation by high-frequency compression and expansion of air bubbles. Therefore, the comparison of the spectral contents of the measured and calculated pressures has only been performed for Fourier coefficients of up to 200 Hz maximum.

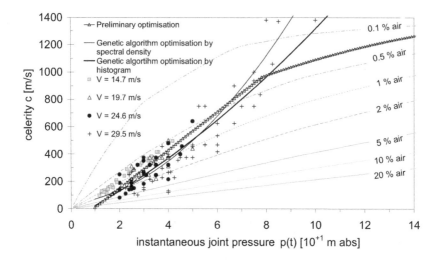

Figure 6. Numerically derived celerity-pressure relationships and comparison with measured data points for different jet velocities and plunge pool depths for turbulent flow assumptions.

5.4 Summary of genetic algorithm optimization

Similar to the preliminary optimisation, Table 5 and Figure 8 summarize the results of the genetic algorithm optimisation for different jet velocities and plunge pool depths. The experimental runs that have been used for the optimisation were for time periods of 10 seconds each.

It can be noted that the friction coefficient λ increases with increasing jet velocity and, based on the corresponding celerity-pressure relationships, also with increasing air concentration in the joint. This coefficient not only represents the classical steady-state friction but also includes all possible two-phase damping effects. It is believed, therefore, that the frictional effect due to the gas phase is by far the predominant one, which is in agreement with previous findings (Martin & Padmanabhan, 1979; Ewing, 1980).

Furthermore, the friction coefficient η, which accounts for unsteady friction effects (or frequency dependent friction), is close to the unity for all tested cases. Figure 5 shows that two types of relationship have been found: quadratic curves, with a slope that increases quite fast with increasing pressure, and linear curves, with a somewhat smaller but constant slope. It has to be mentioned that only the curve for a jet velocity of 24.6 m/s and a plunge pool depth of 0.67 m indicated that a linear relationship fitted better than a quadratic one. The other two linear curves were necessarily obtained because parameter k_3 was omitted from the optimisation, i.e. no possibility of curvature was possible. Within the range of measured pressures, all the curves are in reasonable agreement one with the other. Especially for the curves that have been optimised based on the power spectrum criterion, it can be observed that they are shifted towards higher air contents (lower celerities) at higher jet velocities. Moreover, their shape shows some similarity with the celerity-pressure relationships that are governed by Henry's law.

248

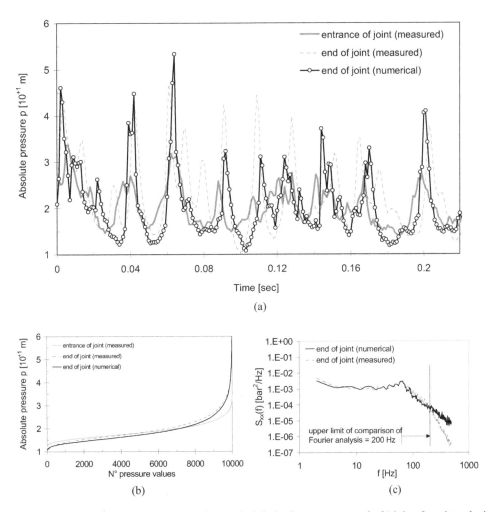

Figure 7. Comparison of experimental and numerical derived pressures at end of I-joint, for a jet velocity of 19.7 m/s and a plunge pool depth of 0.67 m: a) pressure signals in the time domain; b) corresponding histograms; c) corresponding power spectral densities.

Table 5: Parametric results of the genetic algorithm optimisation applied to a 10 seconds test run for different jet velocities and plunge pool depths.

V_j m/s	Y m	k_1 -	k_2 -	k_3 -	c_{PIV} m/s	k_4 -	k_5 -	k_6 -	λ -	η -	crit -
14.7	0.20	65	10.0	0.000	-	-	-	-	0.20	1.00	histo
14.7	0.67	125	9.0	0.071	-	-	-	-	0.35	1.00	histo
14.7	0.67	77.1	12.0	0.021	-	-	-	-	0.47	1.00	spec
19.7	0.67	22	10.2	0.047	-	-	-	-	0.53	1.14	histo
19.7	0.67	70	5.0	0.143	-	-	-	-	0.58	1.00	spec
24.6	0.20	-10.7	0.5	0.173	-	-	-	-	1.00	0.80	spec
24.6	0.67	55.7	11.4	0.000	-	-	-	-	0.71	1.00	histo
24.6	0.67	10.7	6.4	0.147	-	-	-	-	0.78	1.00	spec

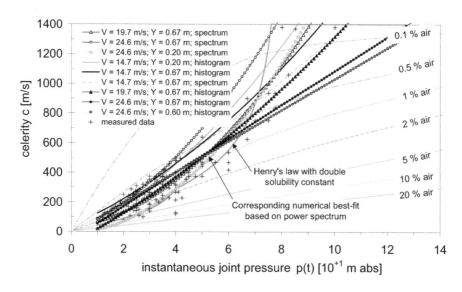

Figure 8. Numerically derived celerity-pressure relationships and comparison with measured data points for different jet velocities and plunge pool depths, for laminar flow assumptions. The numerical optimisation has been performed by a genetic algorithm technique.

6 CONCLUSIONS

The one-dimensional two-phase numerical modelling of the transient pressures that have been measured at the end of the I-shaped closed-end rock joint generates pressures that are in good agreement with the measured ones. The numerical adjustment was based on the optimisation using genetic algorithms of appropriate celerity-pressure relationships. These relationships seem to be governed by the ideal gas law and by Henry's law. However, some significant aspects need to be studied more in detail. The numerical model was unable to account for thermal dissipation effects at high frequencies. These effects are due to subsequent air bubble compression and expansion cycles and increase with frequency. Finally, the numerical modelling should also be performed for open-end joints as well as for different geometrical configurations of the closed-end joints. It is believed that appropriate numerical calibration of the model will be greatly enhanced in the future by the continuously increasing processor speed of computers.

7 REFERENCES

Bollaert, E. 2001. Spectral density modulation of plunge pool bottom pressures inside rock fissures. *Proceedings of the XXIXth IAHR Congress*, Student Paper Competition, Beijing.

Bollaert, E. & Schleiss, A. 2001a. A new approach for better assessment of rock scouring due to high velocity jets at dam spillways. *Proceedings of the 5th ICOLD European Symposium*, Geiranger, Norway.

Bollaert, E., Pirotton, M., Schleiss, A. 2001c. Multiphase transient flow and pressures in rock joints due to high-velocity jet impact: an experimental and numerical approach. *Proceedings of the 3rd International Symposium on Environmental Hydraulics*, Tempe, 3-8 December 2001.

Erpicum, S. (2001) Application des algorithmes génétiques aux problèmes d'optimisation en hydrodynamique de surface, Master Thesis, HACH, University of Liège, Belgium.

EWING, D.J.F. (1980): Allowing for Free Air in Waterhammer Analysis, Proceedings of the 3rd International Conference on Pressure Surges, Canterbury, England.

MARTIN, C.S. & PADMANABHAN, M. (1979): Pressure Pulse Propagation in Two-Component Slug Flow, Transactions of the ASME, Journal of Fluids Engineering, Vol. 101, pp. 44-52.

WHITEMAN, K.J. & PEARSALL, I. S. (1959): Reflux Valve and Surge Tests at Kingston Pumping Station, British Hydromech. Res. Assoc., National Engineering Laboratory joint report N°1.

WYLIE, E.B. & STREETER, V.L. (1978): Fluid transients, McGraw-Hill Inc., US.

History of energy dissipators

La Llosa de Cavall Arch Dam (Spain)
Courtesy of A.J. Schleiss (LCH-EPFL)

Rock Scour due to falling High-velocity Jets - Schleiss & Bollaert (eds)
© *2002 Swets & Zeitlinger, Lisse, ISBN 90 5809 518 5*

Short history of energy dissipators in hydraulic engineering

W.H. Hager
VAW, ETH-Zentrum, Zurich, Switzerland

ABSTRACT: Dissipation of energy in hydraulic engineering is one of the topics for which experimental hydraulics is still essential, given the complications with numerical methods for highly turbulent multiphase flow. The following would like to add to the historical development of hydraulic engineering in terms of energy dissipators and highlight European contributions in particular. After a general introduction to the significance of dissipator structures, methods are outlined that were used in the first decades of the 20th century. At the end of this paper a number of distinguished engineers are presented in terms of short biographies that have considerably added to the present knowledge of energy dissipators.

1 INTRODUCTION

Energy dissipation as a physical process was not understood until the mid's of the 19th century when scientists such as Mayer, Arago and Joule discovered the transformation of mechanical energy into heat, and as a by product also were able to solve the enigma of the *Perpetuum mobile*. Since the middle ages at least, but most probably since ancient times, humans tried to use nature as a resource for work. Provided nature would offer its secrets, science could profit directly and consume free energy, or at least make a process continue eternally once being in motion. This was demonstrated to be impossible, because of energy losses such as by a transfer from, say, hydraulic to potential energies.

Hydraulic engineers are restricted by energy losses, because a turbine would not have an efficiency of 100%, and water in a river has upper limits of velocity due to so-called hydraulic losses. On the other hand; there are processes where a hydraulic engineer would be interested when the hydraulic losses were larger than they actually are. One of this field of application are energy dissipators, i.e. hydraulic structures where a concentrated and significant change of mechanical to other types of energies is welcomed. However, nature also limits these possibilities, and research tried to enhance the transfer from mechanical energy to mainly heat.

It may sound strange that there is still today, in the age where energy should be conserved, a profession which looks for better means to dissipate energy. Anybody 'normal' might just shake head which is to understand from a usual point of view. For hydraulic engineers, safety requirements are severe, because each structure erected should withstand a well-defined hydraulic load. Energy dissipators are particularly important for high dams, given the immense potential of scour of such schemes. To design a dam thus includes also the task to conceive an overflow structure in order to control *all* floods defined by hydrology. There are two particular points that require an overflow structure: (1) Water consumption from any reservoir is based on an average discharge, such as in irrigation techniques or when the reservoir is used for hydraulic power production, and (2) Full reservoirs are more interesting in terms of hydraulic management. Consider a reservoir used exclusively for power production. (1) thus states that the maximum discharge is much larger than the design discharge of the turbines, whereas (2) indicates that the turbines have maximum efficiency for a full reservoir. The overflow structure will thus have to be designed for a completely filled reservoir under the maximum inflow hydrograph by accounting for additional safety factors. It seems clear that for such a scenario, energy dissipation becomes a concern. Consider a typical dam elevation of 100 m. If friction is excluded for the moment, the velocity at the foot of the dam would be of the order of 30 m/s, corresponding to some 100 km/h. Discharges up to several thousands of cubic meters per second are nothing extraordinary and the energy available during such phenomena is just fantastic.

The other point that needs consideration is the downstream site if nothing would be provided to tame the energy released: It is easy to imagine that a valley can easily be destroyed, producing large damages for humans and infrastructure. In addition, because of erosion, slope instabilities and sediment transport, the dam site itself may be highly endangered. Accordingly, provisions must be taken to control the effects of floods across any hydraulic works as a general principle of dam safety. Similar arguments apply to any hydraulic structure, because the loss of one of its members results in a loss of control over a man-made work.

The following paper would like to highlight several features of the history of energy dissipators related to dam engineering mainly. For river engineering, less rigid methods were applied because of normally smaller flow velocities. Nature provided a natural cascade of energy dissipation for rivers which can be demonstrated by the relation between average grain size and slope in mountainous regions. For flatlands, a completely different mechanism involves the formation of river meanders and river braiding, as a local increase of energy dissipation. These methods cannot be applied in dam engineering, however, because they would be inappropriate for the forces described previously.

2 HISTORICAL BACKGROUND

Energy dissipators were not a concern of hydraulic research up to about the 1920ies, when persons like Theodor REHBOCK (1864-1950) started to propose means to control a high-speed flow downstream of a dam. This was closely related to the advance of dam engineering, with dam heights around 30 m around 1900, then increasing some 100 m around 1930, until the maximum heights of almost 300 m in the 1950ies (Schnitter 1994). This technological development was based on the combined advances in material technologies, soil mechanics, geotechnical engineering, concrete improvement and static knowledge, in addition to hydraulic engineering. The need for energy dissipators was a logic result of damages that had occurred wherever foundations were destroyed or other unsatisfactory conditions resulted. Prior to Rehbock, others have successfully tamed the waters downstream of a dam, such as the colonial engineers that were in charge of rivers like the Nile in Egypt, or the Ganges in India. Take as an example the first Aswan barrage erected by John AIRD (1833-1911) as a contractor, and William WILLCOCKS (1852-1932) as the design engineer. Hydraulic heads were around 15 m that produced maximum velocities of some 15 m/s, and which were controlled by a sound kind of stilling basin made of sufficiently large square blocks. It should be remembered that at the time of construction around 1900, no hydraulic model studies were made and that these works have partly been conserved until today.

At that time, prior to scientific hydraulics, three different qualities determined a successful engineer: (1) Sound education, (2) Technical sense and (3) Abstraction. Education was normally provided in specialized institutions such as technical universities, a technical sense could be learned by experience to obtain the right compromise between what was possible and still economical, and the abstraction would allow an engineer to compare his actual project with other successful projects that had similar design characteristics. It was around 1900 when engineering encyclopedias were in fashion, containing a number of recent designs along which a novel project could at least be oriented. It should be remembered that no congresses in dam engineering existed, given that the International Congress of Large Dams ICOLD was founded only in 1930. The exchange in engineering knowledge was relatively small, therefore, such that there were several internationally known dam engineers that would be in charge for projects far away from their country. Some persons of interest are the Germans Otto INTZE (1843-1904) also known for a very early book on dam design and Paul ZIEGLER (1860-1943), the Italian Gaetano GANASSINI (1875-1932), the Spanish José Luis GOMEZ NAVARRO (1869-1954) and Antonio SONIER (1859-1930), the Swiss Fred A. NOETZLI (1887-1933) who was active in the USA and Conradin ZSCHOKKE (1842-1918), and the English William J.E. BINNIE (1867-1949) and Charles HAWKSLEY (1839-1917). Dam heights were limited due to mainly engineering knowledge that developed only in the 1920ies.

3 BASIC TYPES OF ENERGY DISSIPATORS

There are two main structural types by which energy may be dissipated in hydraulic engineering: (1) Hydraulic jump basins and (2) Trajectory basins. For both, the discharge is carried close to the base of the dam and then either directed onto an almost horizontal apron of the hydraulic jump basin, or ejected into the air and directed to a plunge pool. There is a fundamental difference in the mechanism of the two types of basins.

In a *hydraulic jump basin* the energy of the high-speed flow is dissipated by the action of a hydraulic jump, i.e. sufficient tailwater must be available to form the sequent depth required, according to the momentum equation. This requirement may not always be satisfied, because of topographical or hydrological conditions. The efficiency of hydraulic jump basins may be improved with blocks, baffles, sills or other appurtenances that tend to enhance turbulence production such that the basin length can be reduced as compared with the length of a classical hydraulic jump. Problems with these basins are mainly related to high-speed flow that can damage concrete by abrasion, cavitation or up-lift. Also, these concrete basins are relatively expensive because of the immense forces that occur during large floods. Today, hydraulic jump basins are not a common design anymore expect when site conditions do not favor a trajectory basin. However, guidelines are available to design standard basins. These basins are particularly suited for smaller discharges, maybe combined with stepped-spillway designs.

Trajectory basins are a relatively novel design introduced in the 1930ies by the Frenchman Coyne. Instead of directing the high-speed flow horizontally towards the tailwater, the discharge is thrown into the air by a flip bucket and directed away from the dam base. Because of high-speed flow and turbulence associated, such a jet has a large potential to entrain air prior to impinge onto the tailwater. Accordingly, an air-water mixture flow is directed to normally a rocky surface which eventually develops into a plunge pool. Care must be taken that the impact area is safe in terms of retrogressive scour towards the dam or valley sides, under all possible load conditions. Problems with these basins are poor understanding of a highly complex scouring process involving three phases combined with rock cracking and local depositions that may submerge the area concerned with an indirect effect on the flip bucket. Whereas well-defined concrete is used for stilling basins, plunge pool basins involve mainly rock that may be subject to large variation in material properties. Yet, the latter design is currently much in favor to hydraulic jump basins, and is particularly employed for water velocities in excess of, say, 20 m/s. A further reason for using plunge pool basins is the trend to consider mainly embankment dams during the last several decades for which spillways cannot be placed on the downstream dam side. It may be noted as for the stilling basin, that trajectory spillways were hardly model-tested prior to World War II, and only damages in the after war years led to a more detailed hydraulic research of this attracting type of energy dissipator.

4   MODERN HYDRAULIC ENGINEERING

It was stated that the hydraulic engineer had about three tools to approach a technical problem around 1900: Education rendering basic tools for a hydraulic calculation, experience to transmit knowledge from successful previous designs, and abstraction to avoid serious mistakes. Starting in the first decades of the 20th century, a forth even more important aid became available with hydraulic modeling. Around 1900 there were large critics by practical engineers against the analysis of hydraulic phenomena with model tests. It must be stated that similarity laws were then not yet clarified, and these persons were not ready to believe that a small 'toy' would produce conditions of a prototype. Yet, as first demonstrated by Engels and later especially by Rehbock, a hydraulic model is able to reproduce at least the main feats of a hydraulic scheme provided several restrictions are observed. These include for energy dissipators in particular problems with air-water flow, with spray, with fluid viscosity and with sediment. Model tests became reliable in the 1920ies, finally, because of the successes scientists previously mentioned had. In the post-war era, model tests became a must for each hydraulic structure of relevance. In the early 1960ies, a still other aid to predict hydraulic phenomena was introduced with numerical modeling. Yet, energy dissipators are so complex in terms of multiphase flow, turbulence level and turbulent pressure variations that computational fluid dynamics CFD is even currently of no direct aid. That is not to state that CFD has not added to hydraulic engineering in general, yet problems with energy dissipators continue to be investigated mainly by experimental means.

5 DEVELOPMENTS IN THE 1930IES

Germany was able to recover relatively fast from World War I, as may also be noted from the outstanding book European laboratory practice by the American John R. Freeman in 1929. For one or the other reason, the UK was completely absent in this historical book, and France added only marginally. Starting around 1900, three main hydraulic laboratories had taken over leadership in German speaking countries, namely those of the technical universities of Berlin then directed by Adolf LUDIN (1879-1968), and Hans Detlef KREY (1866-1928), of Karlsruhe technical university as already mentioned, and of the Austrian technical university of Brno (today Czech Republic) directed by Armin SCHOKLITSCH (1888-1969). The university of Dresden once known for the first hydraulic lab under the leadership of Hubert ENGELS (1854-1945) had no important role in dam engineering. Energy dissipators were significantly developed at the three institutions mentioned, with questions relating to sediment at Berlin, Rehbock's dentated sill design at Karlsruhe and a general approach by Schoklitsch. Up to World War II, several other institutions followed the successful German school that declined after 1935 because of political changes. Notable persons in relation to energy dissipators were the Austrian Rudolf EHRENBERGER (1882-1956), the Czechs Jan SMETANA (1883-1962) and Antonin SMRCEK (1859-1951), the Frenchmen André COYNE (1891-1960) and Léopold ESCANDE (1902-1980), the Germans Emil MATTERN (1865-1935) and Wilhelm SOLDAN (1872-1933), the Italians Guido NEBBIA (1894-1947), Ettore SCIMEMI (1895-1952), Carlo SEMENZA (1893-1961) and Alessandro VERONESE (1894-1968), the Portuguese Antonio Trigo de MORAIS (1895-1966) and J.F. PINTO REBELO (1907-1979, the Spanish Jose Luis GOMEZ-NAVARRO (1869-1927), the Swiss Heinrich Eduard GRUNER (1873-1947) and Alfred STUCKY (1892-1969) and the English John Guthrie BROWN (1892-1976), Alexander GIBB (1872-1958) and Serge N. LELIAVSKY (1891-1963). Their main activity was before 1950 which limits the survey period of the present paper.

6 PERSONALITIES

To obtain further light on energy dissipators, consider selected personalities that have significantly contributed to the art of taming wild waters. Some of the persons previously mentioned are well documented and no additions are made here, such as for Rehbock, Intze, Noetzli, Zschokke, Ludin, Schoklitsch, Engels, Scimemi and Stucky (Garbrecht 1987, Franke und Kleinschroth 1991, Vischer 2001). Let us thus consider persons that have particularly contributed to the hydraulics of energy dissipators with short biographies. In total, representatives of 10 different European countries have been retained, namely from Austria, the Czech Republic, France, Germany, Italy, Portugal, Russia, Spain, Switzerland, and United Kingdom. The selection is somehow arbitrary but allows an additional overview on the development of energy dissipators. The biographies are accompanied with three main references indicating biographical sources and main works of the person considered. The sign *P* indicates the source to the portrait.

EHRENBERGER

Rudolf Ehrenberger's name remains connected to *Versuchsanstalt für Wasserbau* at Vienna, which was inaugurated in 1912 and where he worked for over thirty years. He was born on August 16 1882 at Krems. Ehrenberger's research included velocity and pressure distributions along drop structures, first observations on air entrainment on chutes, and energy dissipators. He was involved in contract works too, such as for the hydropower plant *Ybbs-Persenbeug* on Danube river. He retired in 1951 and passed away at Vienna on April 27 1956. Ehrenberger demonstrated ability in hydraulic modeling using novel experimental techniques. Together with Schoklitsch he is considered an notable Austrian experimenter that was awarded a title *Ministerialrat* after World War II.

Ehrenberger, R. 1929. Verteilung der Drücke an Wehrrücken. *Wasserwirtschaft* 22(5): 65-72.
Ehrenberger, R. 1930. Eine neue Geschwindigkeitsformel für künstliche Gerinne mit starken Neigungen. *Wasserwirtschaft* 23(28): 573-575; 23(29): 595-598.
Wibmer, K. 1988. 75 Jahre Bundesanstalt für Wasserbauversuche und hydrometrische Prüfung in Wien. *Österreichische Wasserwirtschaft* 40(7/8): 169-175. *P*

SMRCEK

Antonin Smrcek was born on December 10 1859 at Brodek in Moravia, a part of the Austrian empire, graduated in civil engineering at Prague technical university in 1884, and joined in 1888 a consulting office, finally as chief engineer with works for *Elbe* and *Vltava* rivers. He was in 1902 appointed hydraulics professor at the newly erected Brno Czech technical university. In 1917 a hydraulic lab was inaugurated which he directed until retirement in 1932. He died on February 17 1951 at Brno. Smrcek is known for hydraulic works relating to spillway flows. He suggested various designs for energy dissipators and investigated the effect of appurtenances. Smrcek was also one of the first to propose the side channel.

Freeman, J.R. 1929. The hydraulic laboratory of the Bohemian technical university of Brno. *Hydraulic Laboratory Practice*:477-516. ASME: New York.*P*
Pospisil, J. (1960). Vzpominka na prof. Dr.h.c. Smrcka. *Vodni Hospodarstvi* 10(1): 48.*P*
Smrcek, A. (1930). Modellversuche über Einrichtungen zur Regulierung des Überfalls. *Gesamtbericht Weltkraftkonferenz* 2 Berlin 9(393): 353-378.

COYNE

André Coyne, born on February 17 1891 at Paris, made his engineering degree with *Ecole Polytechnique*. In 1920, as a member of *corps des Ponts et Chaussées*, he contributed to harbour works at Brest. Promoted to chief engineer in 1931, Coyne started his long career in dam engineering along Dordogne river, with thin arch dams *Marèges* and *Saint-Etienne-Cantalès*. He was appointed head of dam service in 1935, and promoted to *Inspecteur Général* in 1941. In 1947, Coyne founded a consulting office with Bellier, and collaborated with *Eléctricité de France* EdF. Coyne was consultant for over 100 international dam projects. At *Aigle* dam, the first ski jump spillway was incorporated into the powerhouse. From 1946 to 1952, Coyne presided ICOLD, was a honorary member of ASCE, among many other prestigious awards.

He also designed *Malpasset* dam in Southern France that failed in 1959, and shortly later passed away at Paris on July 21, 1960.

Billoré, J. 1991. André Coyne: A review of his ideas and their application today. *Water Power and Dam Construction* 43(6): 47-54. *P*

Caquot, A., Bernard-Renaud, M., Decelle, A., Mary, M., Laprade, A. 1967. André Coyne. *La Houille Blanche* 22: 135-139. *P*

Coyne, A. 1937. Construction of large dams. *Water and Water Engineering* 39:89-100.

ZIEGLER

Born on August 5 1860 at Gotha, Paul Ziegler made civil engineering studies at Hannover technical university and joined the Prussian hydraulic administration. He was sent to *Ruhrtal* in 1889 where he came in contact with dams. In 1895, after having been active for the *Nordostseekanal*, he returned to *Wuppertal* and was involved with *Bever* dam among other hydraulic structures used for mining. From 1905 onwards he directed the *Harz* association, responsible for flood protection. He designed hydraulic schemes although an association for the *Harz* region was only founded in 1928 because of the first world war. Ziegler authored a book on dam engineering, with typical photographs illustrating the rapid development of early dam design. He died at Clausthal on March 17, 1943.

Franke, P.-G. 1993. P. Ziegler, ein Pionier des Talsperrenbaues. *Wasser und Boden* 45(3): 177. *P*

Ziegler, P. 1900. *Der Thalsperrenbau*. Seydel: Berlin.

Ziegler, P. 1934. Vorbecken, Leerschüsse, Umleitungen und Ausgleichbecken von Talsperren. *Deutsche Wasserwirtschaft* 29(7): 135-138.

NEBBIA

Guido Nebbia, born on May 20 1894 at Campobasso, obtained the civil engineering degree from Napoli university only in 1920 because of war. After having been assistant he gradually climbed the academic ladder until 1942 to take over as hydraulics professor of Napoli university in 1944, and was promoted to president of the civil engineering faculty and vice-rector of that institution. Nebbia largely contributed to classical hydraulics of the Italian school with main interests in hydraulic jumps and stilling basins such as with bottom drops or with transverse sills. He also contributed to scour problems downstream of energy dissipators. Nebbia passed away on April 24 1947 at Napoli due to overwork.

Anonymous 1994. A ricordo di Guido Nebbia. *24 Convegno di Idraulica e Costruzioni Idrauliche* Napoli. CUEN: Napoli, with bibliography. *P*

Nebbia, G. 1940. Sui dissipatori a salto di Bidone. *L'Energia Elettrica* 27(3): 125-138; 27(6): 325-355; 28(7): 441-454; 28(8): 533-546.

Russo, G. 1967. *La scuola d'ingegnere in Napoli 1811-1967*. Istituto Editoriale del Mezzogiorno: Na poli. *P*

MORAIS

Antònio Trigo de Morais was born on February 3 1895 at Vial Flor and graduated as a civil engineer from *Instituto Superior Técnico*, Lisboa in 1918. His career started in 1924 with hydraulic works at *Vale do Buzi*, Mozambique. After having returned to Portugal in 1934 as a member of the public works ministry he was involved in the design of five large dams. In 1949, the agricultural section of the ministry of public works was integrated in the national hydraulics service with Morais as the general director. Later, he returned to Angola and Mozambique, and added to hydraulic schemes of the *Limpopo* and the *Cunenen* rivers. Morais was also involved in the design and execution of almost 20 dams that served mainly for agricultural purposes. He may be considered a pioneer of agricultural hydraulics, therefore. Morais died on February 15 1966 at Lisbon.

Morais, A. Trigo de 1945. *Sempre o problema da agua*. Bertrand: Lisboa.

Morais, A. Trigo de 1951.*A agua na valorizaçao do Ultramar*. Agência Colonia:Lisboa.

Sanches, R. (1995). No centenario do engenheiro Antonio Trigo de Morais. *Ingenium* 10(1/2): 81-88. *P*

GUN'KO

Born on May 6 1911, Fedor Grigor'evich Gun'ko worked during all his life at the Leningrad *Vedeneev* All-Union scientific research institute of hydraulic engineering VNIIG, entering there in 1935, obtaining the degree of candidate of technical sciences in 1946, and in 1956 the Russian PhD title and finally being promoted to director of the laboratory of hydraulics and of hydraulic structures in the 1960ies. Gun'ko was known for his contributions to hydraulic structures in particular. He contributed to complex hydropower plants of the former Soviet Union and was known for his works relating to ski jumps, and to scour due to water jets. Among other distinctions, Gun'ko was awarded the Order of October Revolution. He died on February 17 1990 at Leningrad.

Anonymous 1971. 60th birthday of Fedor Grigor'evich Gun'ko. *Hydrotechnical Construction* 5(8):787. *P*

Gun'ko, F.G., Burkov, A.F., Isachenko, N.B., Rubinstein, G.L., Yuditsky, G.A. 1965. Scour of river bed below spillways of high-head dams. *11 IAHR Congress* Leningrad 1(5): 1-14.

Gun'ko, F.G. 1967. Macroturbulence of flows below spillways of medium head dams and their protection against undermining. *12 IAHR Congress* Fort Collins 2(B16): 135-142.

Gun'ko, F.G., ed. 1974. *Hydraulic calculations of structures controlling rapid flows.* Energiya: Moscow (in Russian).

GOMEZ-NAVARRO

José Luis Gomez-Navarro was born in 1869, graduated in 1927 from Madrid university as a civil engineer and entered *Boetticher* and *Navarro* enterprises, where he was in charge of dam design first, then advanced to director and finally took over as president. His activities were especially directed to *Villaverde* dam where he successfully integrated a ski jump into the spillway. In parallel, Gomez-Navarro was professor of hydraulic structures at Madrid university, and is considered its founder in Spain. His 1944 book looks outstanding and is a summary of hydraulic engineering of that time, treating weirs and spillways, backwater curves and energy dissipators, intake structures and the design of canals, pressurized steady and unsteady pipe flows, and finally chapters on hydraulic modelling and on the ski jump, then a novel method for energy dissipation. He died in 1954.

Anonymous 1950. J. L. Gomez-Navarro. *Revista de Obras Publicas* 98(12): 629-630.

Anonymous 1954. Ha muerto D. José Luis Gomez-Navarro. *Revista de Obras Publicas* 102(6): 261-263. *P*

Gomez Navarro, J.L., Aracil, J.J. 1944. *Saltos de agua y presas de embalse.* Tipografia Artistica: Madrid.

STUCKY

Together with Eugen Meyer-Peter (1883-1969), Alfred Stucky was a Swiss exponent of hydraulic engineering in the golden years of hydropower. Born on March 16 1892 at La Chaux-de Fonds, he obtained the ETH civil engineering diploma in 1915, submitted a PhD thesis to ETH in 1920 on arch dams, and was appointed professor of hydraulic engineering at Lausanne engineering school in 1926. In 1928 he founded the hydraulic lab and also opened a private consulting office in parallel to extensive work at EPUL, in which he transformed the former institution while being president. In 1963, Stucky retired from EPUL and concentrated on consulting until death on September 6 1969 at Lausanne. Stucky was involved with the Grand Dixence dam, still the largest gravity dam world wide. He was awarded the honorary doctorate from ETH in 1955 for outstanding civil engineering.

Anonymous 1969. Alfred Stucky. *Wasser- und Energiewirtschaft* 61(11): 349-350, also *Bulletin Technique de la Suisse Romande* 95: 268-270. *P*

Cosandey, M. 1992. Alfred Stucky - un grand ingénieur et un réalisateur authentique. *Pionniers Suisses de l'économie et de la technique* 10. Etudes d'histoire: Meilen.

Stucky, A. 1962. *Druckwasserschlösser von Wasserkraftanlagen*. Springer: Berlin.

LELIAVSKY

Serge Leliavsky, originally from Kiev, Russia, was born on July 9 1891, educated as engineer at St. Petersburg university in 1915 and built the Wielke Bloto dam, Russia. After the Russian revolution, Leliavsky left for Egypt, to heighten the second Aswan dam. He lectured as professor of irrigation at the Egyptian Royal School of Engineering. From 1927 to 1944 he was chief of the design office, reservoirs and Nile barrage department for Aswan dam. After retirement in 1951, he was a private consultant and passed away at Cairo on July 24 1963. Leliavsky is known as author of the monumental Irrigation and hydraulic design, a standard reference, a book in fluvial hydraulics which originated mainly from experiences in India and Egypt. He was awarded the title Bey from Egypt in 1945, the Crampton Prize in 1938 and the Telford Premium in 1948 from the Institution of Civil Engineers, UK. Leliavsky was an invited lecturer to Imperial college, London, in 1959.

Anonymous 1956. British civil engineers honor ASCE prize winner. *Civil Engineering* 26(10): 689; 26(11): 768. *P*

Leliavsky, S.N. 1955. *An introduction to fluvial hydraulics*. Constable: London.

Leliavsky, S.N. 1955. *Irrigation and hydraulic design*. Chapman & Hall: London.

REFERENCES

Franke, P.-G., Kleinschroth, A. 1991. *Kurzbiographien*, Hydraulik und Wasserbau: Persönlichkeiten aus dem deutschsprachigen Raum. Lipp: München (in German)

Garbrecht, G., ed. (1987). Hydraulics and hydraulic research: A historical review. *50 years IAHR* 1935-1985. A.A. Balkema. Rotterdam

Schnitter, N.J. 1994. *A history of dams*, the useful pyramids. A.A. Balkema: Rotterdam

Vischer, D.L. 2001. *Wasserbauer und Hydrauliker der Schweiz*. Verbandsschrift 63. Schweizerischer Wasserwirtschaftsverband: Baden, Schweiz (in German)

Author index

Annandale, G.W.	63, 153, 187, 201
Arefi, F.	73, 227
Attari, J.	73
Babb, A.F.	33
Belicchi, M.	43
Blatter, A.S.	83
Bollaert, E.	137, 161, 187, 239
Burkholder, D.	33
Canepa, S.	117
Carling, P.A.	83
Caroni, E.	43
Casado, J.M.	55
Castillo, L.G.E.	95
Dittrich, A.	83
Dolz, J.	105
Erpicum, S.	239
Fiorotto, V.	43
Golzari, F.	73
Hager, W.H.	117, 253
Hoffmann, M.	83
Hokenson, R.A.	33
Lopardo, M.C.	55
Lopardo, R.A.	55
Mahzari, M.	227
Mason, P.J.	25
Mathewson, C.C.	175
May, J.H.	175
Melo, J.F.	125
Minor, H.-E.	117
Müller, G.	215
Pirotton, M.	239
Puertas, J.	105
Schleiss, A.J.	3, 161, 187, 227, 239
Wibowo, J.L.	175
Wolters, G.	215

T - #0040 - 101024 - C60 - 254/178/15 [17] - CB - 9789058095183 - Gloss Lamination